MANAGEMENT OF DISUSED RADIOACTIVE LIGHTNING CONDUCTORS AND THEIR ASSOCIATED RADIOACTIVE SOURCES

The following States are Members of the International Atomic Energy Agency:

AFGHANISTAN
ALBANIA
ALGERIA
ANGOLA
ANTIGUA AND BARBUDA
ARGENTINA
ARMENIA
AUSTRALIA
AUSTRIA
AZERBAIJAN
BAHAMAS
BAHRAIN
BANGLADESH
BARBADOS
BELARUS
BELGIUM
BELIZE
BENIN
BOLIVIA, PLURINATIONAL STATE OF
BOSNIA AND HERZEGOVINA
BOTSWANA
BRAZIL
BRUNEI DARUSSALAM
BULGARIA
BURKINA FASO
BURUNDI
CAMBODIA
CAMEROON
CANADA
CENTRAL AFRICAN REPUBLIC
CHAD
CHILE
CHINA
COLOMBIA
COMOROS
CONGO
COSTA RICA
CÔTE D'IVOIRE
CROATIA
CUBA
CYPRUS
CZECH REPUBLIC
DEMOCRATIC REPUBLIC OF THE CONGO
DENMARK
DJIBOUTI
DOMINICA
DOMINICAN REPUBLIC
ECUADOR
EGYPT
EL SALVADOR
ERITREA
ESTONIA
ESWATINI
ETHIOPIA
FIJI
FINLAND
FRANCE
GABON
GEORGIA
GERMANY
GHANA
GREECE
GRENADA
GUATEMALA
GUYANA
HAITI
HOLY SEE
HONDURAS
HUNGARY
ICELAND
INDIA
INDONESIA
IRAN, ISLAMIC REPUBLIC OF
IRAQ
IRELAND
ISRAEL
ITALY
JAMAICA
JAPAN
JORDAN
KAZAKHSTAN
KENYA
KOREA, REPUBLIC OF
KUWAIT
KYRGYZSTAN
LAO PEOPLE'S DEMOCRATIC REPUBLIC
LATVIA
LEBANON
LESOTHO
LIBERIA
LIBYA
LIECHTENSTEIN
LITHUANIA
LUXEMBOURG
MADAGASCAR
MALAWI
MALAYSIA
MALI
MALTA
MARSHALL ISLANDS
MAURITANIA
MAURITIUS
MEXICO
MONACO
MONGOLIA
MONTENEGRO
MOROCCO
MOZAMBIQUE
MYANMAR
NAMIBIA
NEPAL
NETHERLANDS
NEW ZEALAND
NICARAGUA
NIGER
NIGERIA
NORTH MACEDONIA
NORWAY
OMAN
PAKISTAN
PALAU
PANAMA
PAPUA NEW GUINEA
PARAGUAY
PERU
PHILIPPINES
POLAND
PORTUGAL
QATAR
REPUBLIC OF MOLDOVA
ROMANIA
RUSSIAN FEDERATION
RWANDA
SAINT KITTS AND NEVIS
SAINT LUCIA
SAINT VINCENT AND THE GRENADINES
SAMOA
SAN MARINO
SAUDI ARABIA
SENEGAL
SERBIA
SEYCHELLES
SIERRA LEONE
SINGAPORE
SLOVAKIA
SLOVENIA
SOUTH AFRICA
SPAIN
SRI LANKA
SUDAN
SWEDEN
SWITZERLAND
SYRIAN ARAB REPUBLIC
TAJIKISTAN
THAILAND
TOGO
TONGA
TRINIDAD AND TOBAGO
TUNISIA
TÜRKİYE
TURKMENISTAN
UGANDA
UKRAINE
UNITED ARAB EMIRATES
UNITED KINGDOM OF GREAT BRITAIN AND NORTHERN IRELAND
UNITED REPUBLIC OF TANZANIA
UNITED STATES OF AMERICA
URUGUAY
UZBEKISTAN
VANUATU
VENEZUELA, BOLIVARIAN REPUBLIC OF
VIET NAM
YEMEN
ZAMBIA
ZIMBABWE

The Agency's Statute was approved on 23 October 1956 by the Conference on the Statute of the IAEA held at United Nations Headquarters, New York; it entered into force on 29 July 1957. The Headquarters of the Agency are situated in Vienna. Its principal objective is "to accelerate and enlarge the contribution of atomic energy to peace, health and prosperity throughout the world".

IAEA NUCLEAR ENERGY SERIES No. NW-T-1.15

MANAGEMENT OF DISUSED RADIOACTIVE LIGHTNING CONDUCTORS AND THEIR ASSOCIATED RADIOACTIVE SOURCES

INTERNATIONAL ATOMIC ENERGY AGENCY
VIENNA, 2022

Marketing and Sales Unit, Publishing Section
International Atomic Energy Agency
Vienna International Centre
PO Box 100
1400 Vienna, Austria
fax: +43 1 26007 22529
tel.: +43 1 2600 22417
email: sales.publications@iaea.org
www.iaea.org/publications

Printed by the IAEA in Austria
October 2022
STI/PUB/2025

IAEA Library Cataloguing in Publication Data

Names: International Atomic Energy Agency.
Title: Management of disused radioactive lightning conductors and their associated radioactive sources / International Atomic Energy Agency.
Description: Vienna : International Atomic Energy Agency, 2022. | Series: IAEA nuclear energy series, ISSN 1995–7807 ; no. NW-T-1.15 | Includes bibliographical references.
Identifiers: IAEAL 22-01525 | ISBN 978–92–0–134822–7 (paperback : alk. paper) | ISBN 978–92–0–134922–4 (pdf) | ISBN 978–92–0–134722–0 (epub)
Subjects: LCSH: Radioactive waste management. | Radioactive substances. | Radioactive waste disposal. | Lightning conductors.
Classification: UDC 621.039.7 | STI/PUB/2025

FOREWORD

The IAEA's statutory role is to "seek to accelerate and enlarge the contribution of atomic energy to peace, health and prosperity throughout the world". Among other functions, the IAEA is authorized to "foster the exchange of scientific and technical information on peaceful uses of atomic energy". One way this is achieved is through a range of technical publications including the IAEA Nuclear Energy Series.

The IAEA Nuclear Energy Series comprises publications designed to further the use of nuclear technologies in support of sustainable development, to advance nuclear science and technology, catalyse innovation and build capacity to support the existing and expanded use of nuclear power and nuclear science applications. The publications include information covering all policy, technological and management aspects of the definition and implementation of activities involving the peaceful use of nuclear technology. While the guidance provided in IAEA Nuclear Energy Series publications does not constitute Member States' consensus, it has undergone internal peer review and been made available to Member States for comment prior to publication.

The IAEA safety standards establish fundamental principles, requirements and recommendations to ensure nuclear safety and serve as a global reference for protecting people and the environment from harmful effects of ionizing radiation.

When IAEA Nuclear Energy Series publications address safety, it is ensured that the IAEA safety standards are referred to as the current boundary conditions for the application of nuclear technology.

In the early twentieth century it was thought that placing a radioactive source near the end of a lightning conductor would help ionize the near field atmosphere, thereby reducing its electrical resistance and theoretically improving the likelihood that lightning would strike the conductor. Radioactive lightning conductors, also called radioactive lightning rods, terminals, preventers or arresters, have been used for several decades in this manner. It is estimated that hundreds of thousands of these radioactive lightning conductors have been installed worldwide.

No convincing scientific evidence has been produced to show that the radioactive sources in any of the various models of radioactive lightning conductors increase the likelihood of lightning striking them. Over several decades, countries have recognized the need to stop installing these conductors and remove existing devices from the public domain. Many countries have enacted legislation to this effect, yet the problems associated with locating, removing, transporting and conditioning large numbers of radioactive lightning conductor sources have made implementing their removal, safe storage and disposal difficult. A number of Member States have approached the IAEA for support in implementing their national conductor removal programmes.

This publication sets out the actions necessary in removing radioactive lightning conductors from the public domain and managing them safely. The IAEA has compiled and published extensive technical knowledge concerning the management of sealed radioactive sources, which is directly applicable to the management of sources from radioactive lightning conductors. This publication helps Member States assess their particular situation and aids them in developing and implementing policies and strategies for the safe removal of these conductors from the public domain and their subsequent management. It also provides information on their management and that of associated sealed radioactive sources. Issues encountered and lessons identified are also included in this publication.

The IAEA officer responsible for this publication was J.C. Benitez-Navarro of the Division of Nuclear Fuel Cycle and Waste Technology.

EDITORIAL NOTE

CONTENTS

1. INTRODUCTION 1

   1.1. Background 1
   1.2. Objectives 2
   1.3. Scope 2
   1.4. Structure 2

2. TYPES AND CHARACTERISTICS OF RADIOACTIVE LIGHTNING CONDUCTORS 2

   2.1. Historical background 2
   2.2. Overview of radioactive lightning conductors 4
   2.3. Characteristics of radionuclides commonly found in radioactive lightning conductors . 5

3. RADIOACTIVE LIGHTNING CONDUCTORS IN THE PUBLIC DOMAIN: PROBLEMS AND ACTIONS TAKEN 6

   3.1. Problems 6
   3.2. Examples of actions taken by some Member States 9
   3.3. Pathway to effective radioactive lightning conductor management 13

4. NATIONAL PROGRAMME FOR RADIOACTIVE LIGHTNING CONDUCTOR MANAGEMENT 14

   4.1. National policy and strategy 14
   4.2. Programme implementation 15
   4.3. Responsibilities of each party involved 16
   4.4. Training of personnel 18
   4.5. Financing 18
   4.6. Management system 19

5. PREPARATORY WORK (BEFORE RADIOACTIVE LIGHTNING CONDUCTOR REMOVAL) 20

   5.1. Locating radioactive lightning conductor 20
   5.2. Initial inventory verification 21
   5.3. Preparation of site work plan(s) 21
   5.4. Record keeping 22

6. SITE WORK: RADIOACTIVE LIGHTNING CONDUCTOR REMOVAL, HEAD REMOVAL, PACKAGING AND TRANSPORT 24

   6.1. Removal of radioactive lightning conductor 25
   6.2. Packaging and transport of radioactive lightning conductor heads 26
   6.3. Record keeping 26

7. MANAGEMENT OF THE RADIOACTIVE LIGHTNING CONDUCTOR HEADS AT CONDITIONING AND/OR STORAGE SITES 28

7.1. Administrative issues 28
7.2. Initial actions when radioactive lightning conductor heads are received 30
7.3. Treatment/conditioning of sources removed from heads 30
7.4. Record keeping 32

8. CONDITIONING OF THE SOURCES AND SECONDARY WASTE 33

8.1. Selection and qualification of the conditioning method 33
8.2. Equipment and material required 33
8.3. Conditioning of alpha sources 34
8.4. Conditioning of beta/gamma sources 34
8.5. Marking and labelling of packages 35
8.6. Safety and security considerations 35
8.7. Management of secondary waste 35
8.8. Record keeping 36

9. STORAGE AND DISPOSAL OF CONDITIONED SOURCES AND WASTE 37

9.1. Storage 37
9.2. Storage facility 37
9.3. Disposal 38
9.4. Record keeping 39

10. CONCLUSIONS 39

APPENDIX I: EXAMPLES OF RADIOACTIVE LIGHTNING CONDUCTORS 41

APPENDIX II: EXAMPLE OF A PUBLIC AWARENESS PROGRAMME 53

REFERENCES 55

ANNEX I: EXAMPLE OF THE FULL LIFE CYCLE MANAGEMENT OF RLCS REMOVED FROM THE PUBLIC DOMAIN: SPAIN 59

ANNEX II: EXAMPLE OF THE MANAGEMENT OF RLCS REMOVED FROM THE PUBLIC DOMAIN: SINGAPORE 62

ANNEX III: EXAMPLE OF THE MANAGEMENT OF RLCS REMOVED FROM THE PUBLIC DOMAIN: CUBA 65

ANNEX IV: EXAMPLE OF THE MANAGEMENT OF RLCS REMOVED FROM THE PUBLIC DOMAIN: PARAGUAY 70

ANNEX V: EXAMPLE OF THE MANAGEMENT OF RLCS REMOVED FROM THE PUBLIC DOMAIN: FRANCE 74

ANNEX VI: EXAMPLE OF THE MANAGEMENT OF RLCS REMOVED FROM THE PUBLIC DOMAIN: MALAYSIA 75

ANNEX VII: EXAMPLE OF THE MANAGEMENT OF RLCS REMOVED FROM THE PUBLIC DOMAIN: MALTA 82

ABBREVIATIONS 87
CONTRIBUTORS TO DRAFTING AND REVIEW 89
STRUCTURE OF THE IAEA NUCLEAR ENERGY SERIES 90

1. INTRODUCTION

1.1. BACKGROUND

In the early twentieth century, it was proposed that placing a radioactive source near the end of a lightning conductor would help ionize the near field atmosphere, reducing its electrical resistance and theoretically improving the likelihood that lightning would strike the conductor. In the 1960s, radioactive lightning conductors (RLCs)[1] were widely manufactured and initially the radioactive sources in these devices were small — on the order of kBq (μCi) — but later increased to a few hundreds of MBq (mCi) per source. ^{226}Ra and ^{241}Am were the first radionuclides to be used. In later years, and particularly in former Yugoslavia, larger quantities of gamma emitting radionuclides were used for this purpose.

Radioactive lightning conductors, also called radioactive lightning rods, terminals, preventers or arresters, were used for several decades. It is estimated that hundreds of thousands of these RLCs were installed worldwide.

The use of RLCs in most countries began before the development of specific legislation to control the possession and use of radioactive materials. Most countries around the world were not concerned about the installation of RLCs and many did not keep records as to where they were installed. With time, it was determined that all of these devices presented radiation protection risks to the public. In particular, leakage resulting in non-fixed contamination, in the case of the ^{226}Ra and ^{241}Am sources, and unnecessary exposure to ionizing radiation, in the case of the higher level $^{154/152}Eu$ and ^{60}Co sources, were the predominant issues. In countries that were involved in conflict, many RLC sources were stolen, displaced, damaged or lost.

No convincing scientific evidence has been produced to show that the radioactive sources in any of the various models of RLCs increase the likelihood of lightning striking them. For the past several decades, most countries have officially recognized the need to stop installing RLCs and to remove existing devices from the public domain. Many countries have enacted legislation to this effect, yet the problems associated with locating, removing, transporting and conditioning large numbers of RLC sources have made implementing their removal, safe storage and disposal difficult. Many of these countries have also approached the IAEA for support in implementing their national programmes. Thus, there is a need for a comprehensive guide setting out the key components and actions to remove RLCs from the public domain and manage them safely. Fortunately, the IAEA has compiled and published extensive technical knowledge concerning the management of sealed radioactive sources (SRSs), which is directly applicable to the management of sources from RLCs.

SRSs are used worldwide in medicine, agriculture, industry and research, in mobile as well as stationary devices. International Basic Safety Standards, IAEA Safety Standards Series No. GSR Part 3, define a sealed source as a "radioactive source in which the radioactive material is (a) permanently sealed in a capsule or (b) closely bonded and in a solid form" [1].

If a source is no longer needed (e.g. replaced by a different technique) or it becomes unsuitable for its intended application (e.g. the activity becomes too weak, the equipment containing the source malfunctions or becomes obsolete, the source is damaged or leaking) it is considered disused. Disused sealed radioactive sources (DSRSs) are recycled or managed as radioactive waste.

Some IAEA Technical Documents have been published dealing with different aspects of the safe management of DSRSs, but problems specific to RLCs have not yet been addressed [2–7]. The options currently available for the safe management of RLCs and their associated radioactive sources, as part of the waste management strategy within the context of international experience, also need to be addressed. The need for a publication on the management of RLCs was highlighted during IAEA Technical and Consultancy Meetings and meetings of interregional and regional IAEA Technical Cooperation projects.

---

[1] A list of the abbreviations used in the text is given at the end of the publication.

1.2. OBJECTIVES

This publication provides information to help Member States to: (a) determine if they have RLCs in the public domain within their territories; (b) define the scope of the problem; and (c) develop and implement a strategy to remove RLCs from the public domain and manage them as radioactive waste.

The objectives of this publication are to help Member States assess the RLC situation in their territories and help them develop and implement policies and strategies for the safe removal of RLCs from the public domain and their subsequent management.

Guidance provided here, describing good practices, represents expert opinion but does not constitute recommendations made on the basis of a consensus of Member States.

1.3. SCOPE

This publication covers all technical and organizational aspects related to the recovery and dismantling of RLCs and the safe management of their associated radioactive sources, including contamination control and decontamination operations associated with leaking sources.

1.4. STRUCTURE

This publication includes the following concerning RLCs and their associated radioactive sources:

— Describes their historical background, types and characteristics;
— Describes examples of incidents that provide the motivation for discontinuation of their use and their removal from the public domain;
— Describes management principles and approaches to compiling their inventory;
— Provides guidance on their management after the minimum requirements have been met;
— Presents lessons learned from both good and bad practices;
— Summarizes examples of completed recovery and management;
— Presents conclusions.

2. TYPES AND CHARACTERISTICS OF RADIOACTIVE LIGHTNING CONDUCTORS

2.1. HISTORICAL BACKGROUND

Lightning conductors were conceived and developed by Benjamin Franklin in the eighteenth century. A lightning conductor is a simple device — it is a blunt or sharp tipped metal conductor that is typically mounted on top of the roof of a building. Conductors are often ~2 cm in diameter. They are connected to a copper or aluminium wire that may also be ~2 cm in diameter. The wire is connected to a conductive grid buried in the ground nearby. The purpose of lightning conductors is to provide a low resistance path to ground that can be used to conduct the enormous electrical currents that are generated when lightning strikes occur.

While the invention of the lightning conductor by Benjamin Franklin dates to 1760, it was not until 1914 that the Hungarian scientist Béla Szilárd developed the first RLC head. It was intended to increase the radius of a lightning protection zone by ionizing the air near its peak and thereby increase the

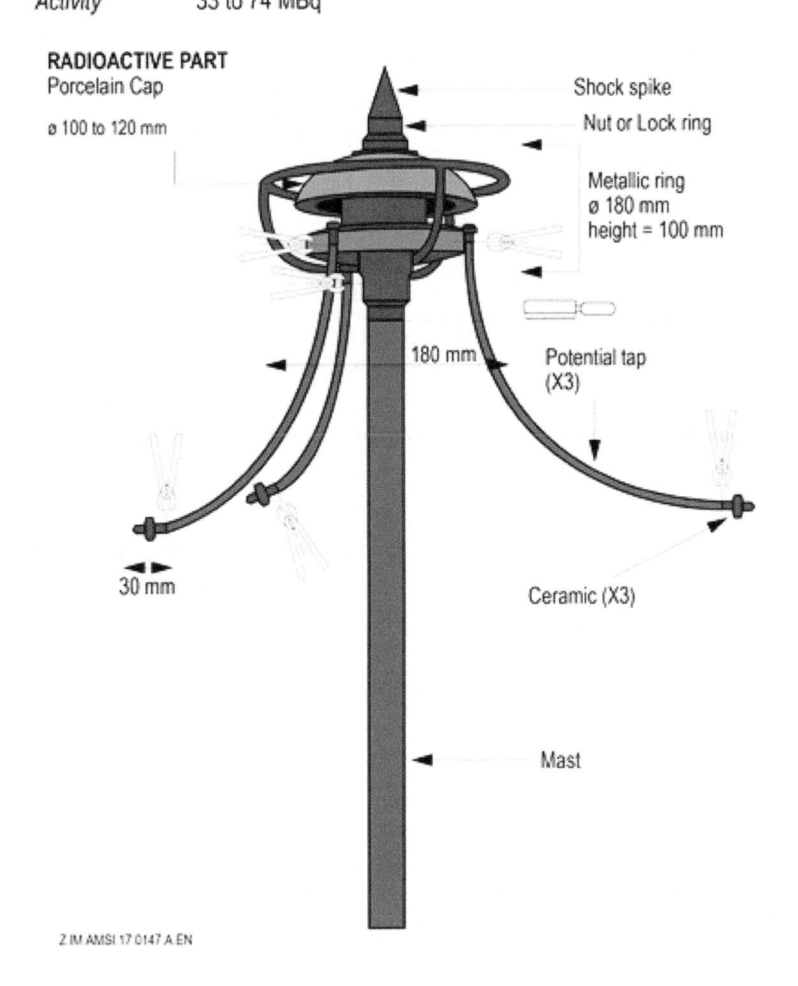

FIG. 1. An early RLC developed by Helita containing a ^{226}Ra source. Courtesy of ANDRA.

probability that lightning would strike the conductor instead of the structure to which it was attached. In 1932, the French company Helita patented the first industrial processes and markets for RLCs (Fig. 1) [8].

The inventors of RLCs claimed that they could attract lightning up to 100 m away, providing a larger protection zone than a conventional Franklin conductor. They promoted the approach that only one centrally located RLC would be required to protect a large building, compared to the dozens of Franklin conductors in a standard lightning protection system.

However, these claims have been repeatedly disproven [9, 10]. The first time was in 1987 when researchers in Australia and Singapore published a study of buildings that had RLCs installed [11]. In the study, several buildings were found to have been struck and damaged by lightning within the claimed protection zones of the RLCs.

Around the same time, RLCs were determined to present a risk to public health, since the radioactive material they contain can disintegrate due to weathering and release radionuclides that can enter the human body via the food chain or through inhaling radioactive dust in the air. In addition, for RLCs with higher activity gamma sources, there is a risk of direct exposure even without the release of radionuclides. Consequently, many countries have banned the manufacture, import, marketing and/or use of RLCs [9].

In the European Union (EU), RLCs are no longer widely available, and so they probably do not represent a significant source of radiation exposure [12].

2.2. OVERVIEW OF RADIOACTIVE LIGHTNING CONDUCTORS

The following summarizes the main features of RLCs.

2.2.1. Device category

Radioactive lightning conductors are categorized according to the radioactive sources they contain, and this is based on the IAEA categorization of SRSs in IAEA Safety Standards Series No. RS-G-1.9, Categorization of Radioactive Sources [13], consisting of:

(a) Mainly Category 5 (most unlikely to be dangerous to the person);
(b) Some Category 4 (unlikely to be dangerous to the person).

Section 2.1 in No. RS-G-1.9 [13] indicates that RLCs were determined to present a risk to public health, yet Categories 4 and 5 in the IAEA categorization of radioactive sources are "**Unlikely to be dangerous to the person**" and "**Most unlikely to be dangerous to the person**", respectively. As a point of clarification, No. RS-G-1.9 [13] also states in Section 2.3 that "At the lower end of the categorization system, sources in Category 5 are the least dangerous; however, even these sources could give rise to doses in excess of the dose limits if not properly controlled, and therefore need to be kept under appropriate regulatory control." The 'risk[2] to public health' relates to damaged or badly managed RLCs, which can lead to a radiation exposure incident, as highlighted in Section 3 of this publication.

There may be situations when many RLCs are collected and stored in the same enclosure. No. RS-G-1.9 [13] addresses the categorization of the aggregation of sources "for the purposes of implementing regulatory control measure" and ought to be referenced in the case of the aggregation of a large quantity of RLCs.

2.2.2. Typical range of dimensions and mass

Radioactive lightning conductors are generally 100–300 mm in diameter and 500–1000 mm in length and weigh 2–10 kg. Appendix I includes images of alpha, beta and gamma type RLCs as well as images of conventional (non-radioactive) lightning conductors for comparison purposes.

2.2.3. Device application

Radioactive lightning conductors were installed to improve the protection of buildings and structures from damage by lightning strikes, based on manufacturers' claims that they provided better protection than conventional non-radioactive lightning conductors.

2.2.4. Functional description

Radioactive sources were attached to lightning conductors to ionize the air near the tips of the conductors.

2.2.5. Typical environment of use

Radioactive lightning conductors were used worldwide, often on buildings that held hazardous materials or sensitive electronic equipment. In some countries they were also installed on many public

[2] Risk is a multiattribute quantity expressing hazard, danger or chance of harmful or injurious consequences associated with exposures or potential exposures. It relates to quantities such as the probability that specific deleterious consequences may arise and the magnitude and character of such consequences.

buildings, such as churches and schools. In the last few decades most countries have operated a programme to remove RLCs from service.

2.2.6. Common sources used in radioactive lightning conductors

Originally, SRS in RLCs contained ^{226}Ra (up to 1.1 GBq (30 mCi) total activity) and ^{241}Am (up to 1.1 GBq (30 mCi) total activity), both of which have relatively long half-lives. These radionuclides are low energy and the SRSs are considered to be Category 5. In later years, and in particular in former Yugoslavia, larger quantities of gamma emitting radionuclides were used for this purpose: typically $^{154/152}$Eu at ~15 GBq (~0.4 Ci) or ^{60}Co at ~7.5 GBq (~0.2 Ci) (Categories 4 and 5). Occasionally, ^{90}Sr, ^{85}Kr and ^{14}C were also used (Category 5) [14]. See Section 2.3 and Appendix I.

2.3. CHARACTERISTICS OF RADIONUCLIDES COMMONLY FOUND IN RADIOACTIVE LIGHTNING CONDUCTORS

The main characteristics of some typical radionuclides used in RLCs are as follows.

2.3.1. Radium

^{226}Ra [15] is part of the radioactive decay series of ^{238}U. It has a long half-life (1600 years) and is a strong alpha emitter with a low level of gamma energy. ^{226}Ra decays by alpha emission to ^{222}Rn, a noble gas with a half-life of 3.6 days, and to four other alpha particles. During decay, many high and low energy gamma photons and beta particles are also emitted. In a radium source, the daughter products are always present, in addition to the parent ^{226}Ra. Therefore, it has a rather high gamma constant.

Old sources of radium present a problem in their management, as they sometimes leak due to mechanical damage or internal over-pressure created by the ^{226}Ra decay, which forms radon and helium gas. The small sizes of the sources prevent marking, and this gives SRSs the deceptively harmless appearance of a small, smooth piece of metal.

Radium is an alkaline earth metal. It is very reactive and even reacts with nitrogen. In radioactive sources, radium is therefore always used in the form of salts, which may be bromides, chlorides, sulphates or carbonates. All are soluble in water in amounts that can give rise to radiological problems. For all these reasons, radium in this form is no longer regarded as a suitable radionuclide for use in SRSs.

^{226}Ra was used in RLCs in the form of ceramic balls/pellets and foils.

2.3.2. Cobalt

Cobalt [5, 16] is a metal element with only one stable isotope, ^{59}Co. When natural cobalt slugs are placed in a nuclear reactor, the nuclei absorb thermal neutrons to make ^{60}Co, a radionuclide with a 5.27 year half-life. ^{60}Co undergoes beta decay (emits an electron and a neutrino) and emits two high energy gamma rays with each decay, finally decaying to the stable isotope ^{60}Ni.

In RLCs, ^{60}Co is usually in the form of thin discs or small cylindrical pellets or slugs welded into stainless steel capsules. Metallic cobalt is not soluble in water and is stable in air, but a thin layer of oxide forms on its surface and this could cause contamination if unprotected cobalt is handled. Incorporation in the body is less dangerous than radium or americium, since it has lower absorption and has higher elimination from the body.

2.3.3. Americium

Americium [5, 16] is an actinide or transuranic element with no stable isotopes. Like the other actinides, americium oxidizes fairly readily. Americium is a decay product derived from ^{238}U activation. ^{241}Am decays with a half-life of 432.7 years by emitting an alpha particle and low energy gamma.

Americium has similar chemical characteristics to rare earth metals. Normally, ^{241}Am is used in oxide form in sealed sources. When used as a low energy gamma source, the stainless steel capsule contains a thin window to allow the gamma photons to be emitted without undue attenuation. As it has higher absorption and lower elimination from the body, intakes can result in a high committed dose.

In RLCs, it is often sealed in a metal foil or deposited and fixed on a ceramic support.

2.3.4. Europium

The europium isotopes in RLCs, which are gamma emitters, are produced by neutron activation in reactors. In RLCs, ^{152}Eu (half-life of 13 years) and ^{154}Eu (half-life of 8.8 years) are typically in powder form in stainless steel capsules. Incorporation in the body is less dangerous than radium or americium, since europium has lower absorption and has higher elimination from the body.

3. RADIOACTIVE LIGHTNING CONDUCTORS IN THE PUBLIC DOMAIN: PROBLEMS AND ACTIONS TAKEN

3.1. PROBLEMS

The problems associated with RLCs are not related to their intended use to promote lightning strikes, but mainly (a) contamination of the site where they were installed; (b) theft, loss or becoming orphaned; and (c) improper management.

3.1.1. Contamination

As RLCs have been in place for many years, generally without maintenance and exposed to all types of weather conditions, it is probable that they have become oxidized and corroded. Oxidation may allow the radionuclide(s) used to escape into the near field environment, resulting in contamination, which may pose risks of exposure to ionizing radiation to workers responsible for removing and dismantling them or to the public.

The physical and chemical characteristics of the radionuclides used may affect the degree of leakage from a damaged source, for example an encapsulated powder versus a metal foil.

It is also possible for sources in an RLC to be damaged and either leak or possibly be dispersed, and such dispersion of material could contaminate other material in the vicinity. Such contamination might be detected, measured and removed using appropriate investigative techniques and procedures.

3.1.2. Loss of control

Structures holding RLCs can decay and deteriorate (Fig. 2), and they can also be stolen or removed and thrown away or sold as scrap metal, with or without the consent of the owners of the structures where they are installed. In either case, RLCs become orphaned devices that are unlikely to be identified as having radioactive materials. IAEA Safety Standards Series No. SSG-17 [17] is concerned with orphan sources and other radioactive material that may enter the metal recycling supply chain.

FIG. 2. Example of a deteriorated structure with an RLC. Courtesy of the Department of Labour Inspection, Cyprus.

Since the use of RLCs started prior to the establishment of an effective regulatory infrastructure in many countries, many are likely to become orphaned. The creation of a national RLC inventory is an important task.

3.1.3. Improper management

The improper management of RLCs often started at the time they were installed. In many cases, their installation was unregulated or incomplete or no lists were created to record where they were installed, and inspection and/or maintenance schedules were not defined.

In many cases, workers maintaining a structure with an RLC installed do not know that these devices contain radioactive material. As a result, workers may remove and handle such devices without knowledge of the risks involved. In other cases, personnel involved in the maintenance or removal of the RLC are not properly qualified and trained for this task, which may lead to their exposure to ionizing radiation or the spread of contamination.

Similarly, persons installing, moving or maintaining lightning conductors, or who are otherwise involved in activities where the lightning conductor assemblies have to be handled, can inadvertently damage one or more of the sources, making them vulnerable to possible leakage of radioactive material.

Solutions to the problem have to be based on radiation protection principles, notably justification for their use and optimization (ALARA[3]). The rationale is that RLCs are no more effective than conventional lightning conductors and thus their use is not justified.

Some incidents involving RLCs are related below.

---

[3] The process of determining what level of protection and safety would result in the magnitude of individual doses, the number of individuals (workers and members of the public) subject to exposure and the likelihood of exposure being "as low as reasonably achievable, economic and social factors being taken into account" [1].

3.1.3.1. Incident involving an radioactive lightning conductor with 152Eu/154Eu sources — Croatia (2005)

Even though RLCs have not been manufactured for years, many are still in regular use in Croatia and they have caused a number of specific radiation protection and regulatory problems [18]. These devices were classified as consumer products, which are usually covered by the general licence concept, with little or no regulatory control. The general licence concept enables persons with minimal or no training in radiation safety to possess and use licensed radioactive sources or devices. The RLCs installed in Croatia used ^{60}Co or $^{152}Eu/^{154}Eu$ with activities of 10–20 GBq, which excluded them from the consumer products category. They are treated as SRSs. Over time, warning labels and signs on RLCs often became obliterated as a result of exposure to adverse environments and improper maintenance. In addition, personnel who are knowledgeable about RLCs retire, are discharged or otherwise leave the licensee's organization. As a consequence of these developments and the absence of control and inspection, some of these RLCs entered the public domain, most frequently by being discarded with scrap metal.

In August 2005, two radioactive sources in their original lead container (open on the upper side) were dismantled from RLCs installed on the roof of one hotel and sold as scrap metal. During transfer of the devices, two 15 GBq $^{152}Eu/^{154}Eu$ sources were dislodged from the containers. The scrap metal was exported to Italy. It was transported from Croatia via Slovenia to Italy by truck. Based on procedures stipulated in Italian law, radioactive material was detected at the Italian border during routine monitoring of the cargo on the truck transporting this scrap metal. Two intact SRSs were discovered (1 cm × 1.5 cm). No leakage of radioactive content was detected. The sources were placed in interim storage and after some time with the consent of the regulatory authority of Croatia they were returned to Croatia and stored in the recognized storage facility. The State Office for Radiation Protection asked for an investigation, dose reconstruction and evaluation of the consequences. Measurements of the source gave 5 mSv/h maximum at a distance of 10 cm and 1 mSv/h at a distance 3.1f 1 m. The driver and passenger were exposed to the sources, as were workers who were handling the scrap metal in the scrapyard. However, only the driver and his companion were identified. They had no radiation monitor during transport. It was estimated that the dose at the driver's position was 100 μSv/h. Based on the transport time, average distance and simulation of the transport, it was estimated that they received an effective dose of ~3 mSv. They were sent for a medical examination, including chromosomal aberration analysis.

3.1.3.2. Incident involving an radioactive lightning conductor with a ^{226}Ra source — Portugal (2011)

On 7 April 2011, the Nuclear and Technological Institute (ITN), the institution responsible for the storage of radioactive waste in Portugal, was contacted by a private company that requested instructions on how to dismantle, handle and transport an RLC containing a Category 5 SRS [19]. The ITN provided the company with detailed instructions. On 4 October 2011, the RLC was finally delivered to the ITN. Prompt inspection and measurements revealed that it contained a ^{226}Ra source; the measured dose rate at contact was 590 μSv/h and at 1 m 12 μSv/h.

After proper investigation, it was determined that the RLC had been dismantled and disassembled one month before delivery at the ITN. The radioactive source had been handled with complete disregard for the instructions provided by the ITN and had been stored for approximately a month underneath an administrative assistant's desk without any shielding. An estimation of the dose received by the administrative staff (one person) exposed to the radioactive source indicated a dose of 1.8 mSv to the torso and one of 7 mSv to the legs. The national statutory annual effective dose limit to a member of the public is 1 mSv. As an additional measure, health examinations of the exposed member of the public would be requested by the occupational health authorities.

3.1.3.3. Incident involving an radioactive lightning conductor estimated to have a $^{152}Eu/^{154}Eu$ source — Serbia (2007)

Upon receiving a written request related to damage of some film tapes, a radiation dose survey was performed at the premises of a film company in Belgrade on 30 May 2007 [20]. Initially, a dose survey was performed in the basement where the films were stored. As assumed by film manufacturer, the observed damage to the tapes was caused by exposure to ionizing radiation. A dosimetric survey in the basement did not result in dose levels higher than the natural background. However, additional survey of the parking lot used for vehicles that transported equipment and crew members showed an increased dose level in the vicinity of a rented truck, which had been used by the team since 13 March 2007. According to truck owner's statement, the truck was bought from another trade company in November 2006. The owner did not have any knowledge about the presence of a radioactive source in the truck.

Accordingly, further investigation was carried out. Detailed dosimetric survey and source recovery was carried out. The reason for increased dose levels was an orphaned source originating from a lightning conductor. On the basis of previous experiences, it was assumed that the source contained a mixture of the radioactive isotopes ^{152}Eu and ^{154}Eu. The dose received by individuals was assessed using a retrospective dosimetry technique based on the information on behaviour of individuals and the results of dose rate measurements in the vicinity of the source. Several people were exposed to relatively high dose rates for an extended period of time. The conservatively estimated cumulative dose values for two categories of individuals were 50 and 40 mSv, respectively, significantly higher than annual dose limit for public exposure of 1 mSv.

Control, accountability, source removal and appropriate disposition of the SRSs within RLCs, along with appropriate record keeping, were recognized as activities that would reduce the probability of radiation incidents occurring in Serbia.

Solutions to the problem have included some or all of: (a) discontinuation and/or banning of their use; (b) removal of these devices from the public domain; and (c) subsequent management of their sources as radioactive waste.

3.1.3.4. Incident involving an ^{241}Am source — Finland (2018)

There were three incidents at the Outokumpu steel mill in Tornio, Finland, between July and September 2018, where a radioactive americium source ended up in melting with recycled metal. A month later (October 2018) a fourth incident took place. This time, however, the americium source was discovered before melting. The alarm was caused by a lightning conductor made in Brazil with three ^{241}Am sources. Nobody was exposed.

The factory has screening detectors for incoming scrap. Despite the modern screening technology used, the source was able to pass through because of americium's low gamma energy, which is easily attenuated by other scrap.

3.2. EXAMPLES OF ACTIONS TAKEN BY SOME MEMBER STATES

Section 3.2 provides examples of some of the actions taken by various Member States to address the problems associated with RLCs in the public domain within their territories. The annexes provide additional, detailed examples of actions taken by some Member States.

Annex I provides an overview of the full life cycle management of RLCs in Spain. It provides an example of what can be done when a Member State has a comprehensive infrastructure to deal with the problem of RLCs in the public domain within its territory.

Annex II provides a brief overview of the management of RLCs in Singapore, which has limited infrastructure and therefore engaged an organization from a Member State that has a comprehensive radioactive waste management infrastructure. Singapore based organizations remove the RLCs from the public domain and dismantled them. The Korean Atomic Energy Research Institute conditioned their sources for storage. Annex III provides a brief overview of the full life cycle management of RLCs in Cuba. Annex IV illustrates a case of meeting the minimum requirements for RLC management in Paraguay. Annex V provides a brief overview of the management of RLCs in France. Annex VI provides a brief overview of the management of RLCs in Malaysia. Annex VII provides a brief overview of the management of RLCs in Malta.

Any decision to remove RLCs from the public domain ought to be based on two principles of radiation protection: the principle of justification [1, 21] — that is, no major benefit is obtained with the use of RLCs — and the principle of optimization (or the ALARA criteria) in order to minimize the risk of exposure of contamination for humans and the environment. Radiological incidents involving RLCs are a strong motivator for removing them from the public domain.

Some European countries produced RLCs in the past and most imported them. RLCs are now prohibited products in most of these countries (Italy, Ireland, Netherlands, Portugal, Spain, Türkiye) [12]. In Luxemburg, lightning conductors containing radioactive sources are no longer available, but some 30 to 40 years ago, many were owned by members of the public. The disposal of this radioactive material is controlled. In Cyprus most RLCs have been removed from buildings and are kept in storage. In the Czech Republic some old RLCs might still be in the public domain. However, RLCs are no longer available for the public to purchase. In Lithuania RLCs are subject to licensing if the dose rate at 0.1 m from the accessible surfaces exceeds 1 μSv/h. In Switzerland RLCs containing ^{226}Ra are prohibited products. Some old ones were installed in the past and are being collected for controlled disposal [12].

3.2.1. Belgium

Radioactive lightning conductors were widely installed and used in Belgium up to August 1985, when a law (Royal Decree) prohibited the installation of additional RLCs. Preliminary estimates made at that time indicated that up to 15 000 RLCs had been installed, with ^{226}Ra, ^{241}Am and ^{85}Kr as the main isotopes.

The removal of the RLCs and their collection by the Belgian National Agency for Management of Radioactive Waste started at that time on a voluntary basis.

In 1997, the Federal Agency for Nuclear Control (FANC) was created by law to take over the former roles of the various bodies responsible for nuclear safety and control in Belgium. The FANC became fully operational in 2001, after the publication of an updated and integrated version of the regulation dealing with the protection of the public, workers and environment against the dangers of ionizing radiation (Royal Decree, 20 July 2001). In those regulations, the use of RLCs was prohibited, with the exception of RLCs installed before 27 October 1985, under the condition that they be periodically controlled by an authorized control organization and found to be in good condition.

In 2003, the FANC, in collaboration with the other organizations involved in the problem, launched a national campaign aiming at solving the problem of RLCs.

The campaign comprised essentially three steps:

(a) Establishing an inventory of places/organizations suspected of having RLCs, essentially on the basis of databases existing in the provinces, municipalities, etc.;
(b) Systematic control of those places/organizations;
(c) Removal of the RLCs by specialized and authorized companies, followed by their dismantling and the collection and further management of the resulting radioactive waste.

For this campaign, the FANC asked for the collaboration of the regional and local authorities. The campaign was also accompanied by public awareness actions in the media (see the example in Appendix II). This campaign also gave rise to a number of parliamentary questions and debates.

At the end of 2010, when it appeared that the situation concerning RLCs in Belgium had been clarified significantly, and that the greatest part of the known RLCs had been removed, the FANC estimated that additional efforts were no longer justified and the active campaign was stopped. However, the programmes that were put in place remain to deal with the few RLCs that may still exist.

3.2.2. Brazil

Brazil experienced a strong market for RLCs in the 1970s and 1980s [22]. These devices typically used ^{241}Am sources, although ^{226}Ra was also used on a smaller scale. The lack of evidence for their superior efficiency compared to the traditional Franklin lightning conductors and the associated risks related to the high radiotoxicity of ^{241}Am were recognized. In 1989, the Brazilian National Nuclear Energy Commission (CNEN) decided that there was no justification for their use and withdrew licences for the manufacture, sale and installation of RLCs. This decision, however, did not imply that already installed devices would be removed.

Many authorities at the municipal level have taken action to replace all installed RLCs under their jurisdiction. Some decisions were taken in the years following the CNEN's decision; in other cities, authorities have only recently awakened to the problem. A good example of effective action is the city of Rio Claro, State of São Paulo, with a population ~180 000 people, where 52 RLCs were removed. In this case, a campaign under the supervision of the Municipal Commission of Civil Defence engaged the local media and service providers and succeeded in mapping, removing and transferring the RLCs to the CNEN.

It was estimated that 75 000 RLCs with ^{241}Am might have been assembled, based on the total activities of ^{241}Am imported. That number may be much lower, however, since, in fact, the CNEN collected a large number of unused ribbons of ^{241}Am from the premises of the manufacturers just after the licence withdrawal in 1989. Up until 2015, approximately 20 000 RLCs have been collected and stored in centralized radioactive waste storage facilities for treatment as radioactive waste. The RLCs are being disassembled inside specially designed glove boxes and the pieces are being stored for further decontamination and recycling. The sources are being stored and will be disposed of together with other DSRSs [23].

3.2.3. Cyprus

In Cyprus, RLCs were supplied and installed by an engineering company in the mid-1960s. It is estimated that ~400 RLCs, using ^{241}Am or ^{226}Ra, and in a few cases ^{85}Kr or $^{152}Eu/^{154}Eu$, were installed on various public or private buildings or other structures in the country. Recently, public concerns were raised concerning the justification and safety of these devices in schools and other public buildings and, as a result, a campaign was initiated for their removal and replacement. All RLCs have been removed from public schools and a great number from private buildings and other structures. The removed RLCs have been collected and dismantled and the radioactive parts are in storage, which is licensed and regularly inspected by the regulatory authority. The conditioning of the collected RLCs was carried out by qualified and trained personnel. At present, the import and installation of new RLCs in Cyprus, as consumer products, is prohibited.

The Radiation Inspection and Control Service has been in contact with the company that imported and installed RLCs in Cyprus and has collected information to prepare an inventory of all RLCs in the country. Furthermore, the regulatory authority has prepared and implemented a comprehensive programme for site inspections to assess the existing situation, provide information and help increase awareness concerning the remaining RLCs. The Radiation Inspection and Control Service aims to replace and dispose of all RLCs in the country.

3.2.4. France

The French National Agency for Radioactive Waste Management (ANDRA) is responsible for managing all radium bearing items by collecting them upon the holder's request [24]. ANDRA is also in charge of collecting RLCs that had been installed, even in individual homes or premises, between 1932 and 1986. Since 1 January 1987, it is forbidden to use any radioactive element in the manufacturing, marketing and import of lightning conductors in France, and according to the Order of 11 October 1983, any dismounted lightning conductor is considered to be radioactive waste. Under standard conditions of use, they do not represent any hazard; but they are still required to be delivered to ANDRA once they are out of service.

3.2.5. Luxembourg

In Luxembourg, in 1995 the Department of Radiation Protection started a programme to withdraw all RLCs in use [25]. These RLCs were installed in the 1960s and 1970s without licences being required. In early 2008, the Department of Radiation Protection contacted all responsible organizations to encourage the removal of the five remaining RLCs. On 23 September 2011, the last of these known RLCs had been dismantled and transferred to the National Interim Storage Facility for radioactive waste.

3.2.6. Malaysia

RLCs were introduced in Malaysia in the 1970s [9]. The use of RLCs in Malaysia is controlled by the regulatory body, where owners have to acquire a licence to own and install RLCs at their premises. However, the import and sale of RLCs were banned in 1989. This was followed by the eventual discontinuation of usage and the removal of RLCs from buildings and premises. Some of the licence owners have on their own initiative removed the RLCs from their installations and disposed of them at the Malaysian Nuclear Agency at their own expense.

At the end of 2017, nearly 150 units of RLCs were received and stored at the Malaysian Nuclear Agency. The efforts by licence owners to remove their RLCs continue at a slow rate, with only one or two more units being removed and disposed of per year between 2018 and 2020. It is believed that some 500 units of RLCs are still available in the public domain throughout the country. Judging from this slow development, the Atomic Energy Licensing Board is currently conducting a survey to update the RLC database where the quantity, location and current conditions of the RLCs are determined and confirmed.

It is planned that the remaining RLCs in the public domain will be removed and disposed of in a national programme with the costs borne by the government. This effort is in line with the Government's objective to protect the safety of the public and the environment from the radiation risks associated with RLCs.

3.2.7. Montenegro

Hundreds of RLCs with ^{152}Eu/^{154}Eu, ^{60}Co and ^{241}Am were estimated to be in Montenegro, situated at various locations, such as schools, public buildings, health establishments and factories. Due to lack of proper maintenance, these RLCs are a source of concern for the public, as some of them have recently been found in scrap metal. The Environmental Protection Agency established an RLC database based on field inspections. Not all RLC locations may be known and the number of installed RLCs might in fact be higher.

The Law on Ionizing Radiation Protection and Radiation Safety (Official Gazette of MNE, Nos 56/09, 58/09) prohibited installation of RLCs in Montenegro. This law prescribes that legal persons and entrepreneurs owning or using RLCs are obligated to remove them. However, financial restraints have impeded progress.

The Ministry of Sustainable Development and Tourism applied for EU Instrument for Pre-Accession Assistance (IPA) funds in order to obtain financial and expert support. The European Commission approved regional project IPA 2009/021-640, 'Management of sealed disused radioactive sources, the removal of radioactive lightning conductors and strengthening the effectiveness of the regulatory infrastructure in the field of radiation in Montenegro, Macedonia and Kosovo (under UNSCR 1244/99)' (project value €1 350 000).

(a) Phase 1: review of the regulatory requirements;

(b) Phase 2: enhancement of capacities within the regulatory authorities and the operator of the Centre for Ecotoxicological Testing. This phase focused on the development of a safety report, working procedures and training courses related to radioactive waste management;
(c) Phase 3: supply of necessary equipment;
(d) Phase 4: removal, transport dismantling and storage of RLCs and other DSRSs from 18 temporary storage facilities.

3.2.8. Spain

In 1986, the manufacture of lightning rods with radioactive elements was prohibited in Spain. Legislation established that owners of RLCs would either officially declare the sites where their RLCs are installed as radioactive facilities, if they wished to keep them, or request their disassembly by Empresa Nacional de Residuos Radiactivos, S.A. [26]. As of 2005, more than 22 000 RLC heads had been removed, most of them containing ^{241}Am. Following their disassembly by the Center for Energy, Environmental and Technological Research (CIEMAT), the recovered disused radioactive sources were sent to the United Kingdom for final management. Although the campaign continues to be open, at present very few of these items still exist, with these being gradually removed and disassembled.

3.2.9. Türkiye

In Türkiye, two types of RLC, ^{226}Ra and ^{241}Am, have been in use since the beginning of the 1970s [27]. The Turkish Atomic Energy Authority (TAEK) prohibited the production, installation and use of radium bearing RLC per letter number 10700-1485 dated 30 July 2001. It is also compulsory that these RLCs are dismounted by agents authorized by TAEK and submitted to the Cekmece Nuclear Atomic Energy Centre. For americium bearing RLCs, only their production has been stopped. However, by taking into consideration the danger that the palladium that covers the ^{241}Am source may oxidize and expose the ^{241}Am, leading to contamination, it is intended that these ^{241}Am RLCs be dismounted by TAEK within 10 years of their installation. It is compulsory that existing ^{241}Am bearing RLCs are checked at least once every year.

3.2.10. Greece

Lightning rods containing ^{226}Ra and ^{241}Am were used in Greece from 1975 until 1985. In these types of rods, the radioactive material is embedded in the head element and not encapsulated in sealed form. This arrangement imposes restrictions on the safe management of the radioactive material.

In the past, the Greek Atomic Energy Commission (EEAE) implemented a national project for the identification, collection and export of disused and orphan sources. Within this project approximately 2000 lightning rods were exported for recycling at an authorized facility.

It is estimated that fewer than 1000 lightning rods are still in place. The EEAE encourages the replacement of the rods and may provide guidance or perform radiation protection measurements if asked by users. To date, two private companies have been authorized in Greece to perform decommissioning and temporary storage of the radioactive lighting rods, upon request. Approximately 500 radioactive lightning rod heads are stored at these facilities, pending final management.

The import and installation of radioactive lightning rods no longer take place.

3.3. PATHWAY TO EFFECTIVE RADIOACTIVE LIGHTNING CONDUCTOR MANAGEMENT

For Member States with fully developed regulatory and waste management infrastructures, dealing with the problem of RLCs in the public domain is a relatively straightforward matter when compared to

Member States with missing or limited infrastructures. The minimum requirements include determining the scope of the problem (create an inventory of RLCs in the public domain) and removing RLCs and placing them in storage. With the exception of locating RLCs, which can pose significant challenges (see Section 5), these activities can be readily achieved, subject to funding and available human resources. Likewise, the subsequent life cycle management of RLCs removed from the public domain can be readily achieved, again subject to funding and available human resources.

In recognition that many Member States may not have the infrastructure to address the problem of RLCs within their territory, this publication provides guidance on implementing a suitable regulatory and management infrastructure (Section 4), as well as guidance on the:

(a) Location of RLCs and preparation of a national inventory (Section 5);
(b) Removal and dismantling of RLCs (Section 6);
(c) Management of RLC heads (Section 7);
(d) Conditioning of sources removed from RLC heads (Section 8);
(e) Storage and/or disposal of conditioned sources (Section 9).

4. NATIONAL PROGRAMME FOR RADIOACTIVE LIGHTNING CONDUCTOR MANAGEMENT

The IAEA has issued safety standards and security guidance publications emphasizing the necessity of national programmes to ensure the safety and security of DSRSs [28–31], including those from RLCs.

Following the requirements and guidance given in those publications relevant to the management of DSRS in general, the national RLC management programme will include the following main elements:

(a) A national policy and strategy (Section 4.1);
(b) Implementation of national programmes (Section 4.2);
(c) Identification of all parties involved in the various steps of RLC removal and management and specification of their responsibilities (Section 4.3);
(d) Provision of resources — financing, technical capability, personnel, personnel qualification and training (Sections 4.4 and 4.5);
(e) A management system (Section 4.6);
(f) Identification of existing RLCs (i.e. a national register) (Section 5);
(g) A public information programme (Section 5.1 and Appendix II).

4.1. NATIONAL POLICY AND STRATEGY

To ensure the effective management and control of radioactive waste, the government is to ensure that a national policy and a strategy for radioactive waste management are established [30], including a policy and strategy for RLC management. The national policy will be based on general principles, such as:

(a) The generation of radioactive waste has to be kept to the minimum practicable level by means of appropriate design measures and procedures, such as the recycling and reuse of material, IAEA Safety Standards Series No. SF-1, Fundamental Safety Principles [28]. The generation of radioactive waste during decommissioning of RLCs will be kept to the minimum reasonably practicable in terms both of activity and volume.

(b) Radioactive sources and waste will be managed safely, including in the long term.
(c) The cost of the management of radioactive waste from RLC will preferably be borne by those who own RLCs; if the owner is unknown or unable to pay, the cost will be borne by the government or another entity (e.g. a fund raised by SRS users for the purpose of funding DSRS management). As stated in IAEA Safety Standards Series No. GSR Part 1 (Rev. 1), Governmental, Legal and Regulatory Framework for Safety [29] (requirement 10), "**the government shall make provision for the safe decommissioning of facilities, the safe management and disposal of radioactive waste arising from facilities and activities, and the safe management of spent fuel**".
(d) A risk based and documented decision making process will be applied with regard to all stages of the removal of RLC from the public domain and subsequent management.

National strategies to implement those policies include the following:

(a) The overall objectives;
(b) The significant milestones and clear time frames for the achievement of those milestones [32];
(c) An inventory of all RLCs to be managed (Section 5.2);
(d) The concepts or plans and technical solutions for radioactive waste management from generation to disposal;
(e) The research, development and demonstration activities that are needed in order to implement solutions for the management of radioactive waste from RLCs;
(f) The responsibility for the implementation of the national programme and the key performance indicators to monitor progress towards implementation [32];
(g) An assessment of the national programme costs and the underlying basis and hypotheses for that assessment, including budget allocations over time;
(h) The financing scheme;
(i) The agreement(s), if any, concluded with another country on the management of specific radioactive waste, including waste from RLCs.

An example of a national strategy for the management of RLCs in a country with a reasonably well advanced infrastructure is provided in Fig. 3.

An example of a national strategy for the management of RLCs in a country with a less advanced nuclear programme and infrastructure is provided in Fig. 4.

4.2. PROGRAMME IMPLEMENTATION

The removal of RLCs from the public domain has to be performed by suitably qualified and experienced personnel (SQEP) who have been trained in the specific procedures to be followed. Some aspects of the work, such as taking an RLC down, only involve conventional work safety (e.g. working at heights) and other aspects involve radiation protection. In general, work could be conducted by a contractor:

(a) Qualified for both conventional work safety and radiation protection; or
(b) Qualified for conventional work safety and supervised by a radiation protection specialist.

A detailed work plan, which describes the method of work and the various steps to safely conduct it, needs to be prepared by the organization that will carry out the removal and management of RLCs. Authorization for removal, transportation, conditioning and storage of the sources is generally needed in advance.

4.3. RESPONSIBILITIES OF EACH PARTY INVOLVED

The relevant principal parties and other parties having specified responsibilities in relation to protection and safety are to ensure that all personnel engaged in activities relevant to protection and safety have appropriate education, training and qualification so that they understand their responsibilities and can perform their duties competently, with appropriate judgement and in accordance with procedures [1].

Responsibilities for each aspect of the work plan need to be clearly stated. Such aspects include the initial contact of the property owner and the removal, transportation, conditioning and long term

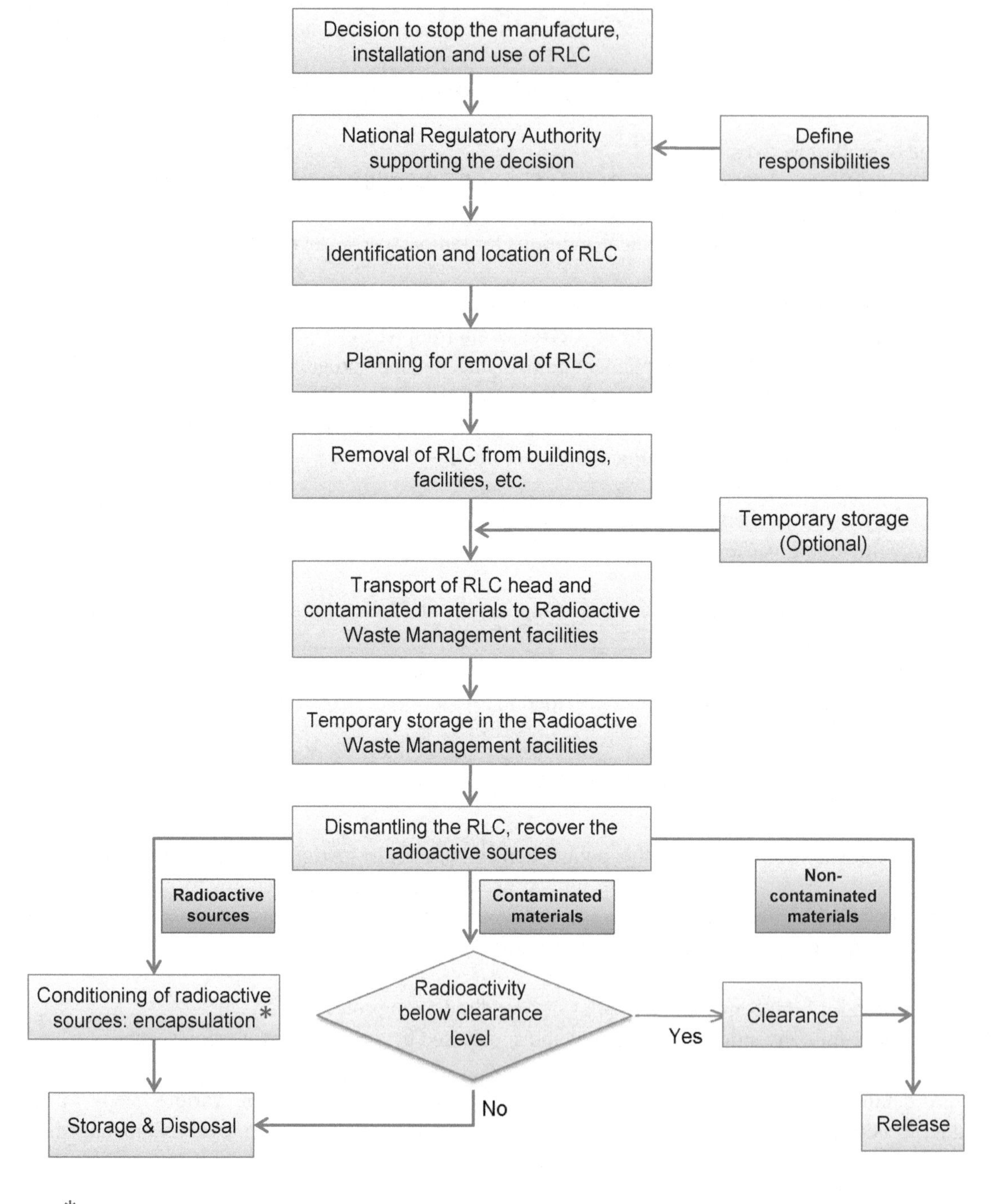

*In the case of ^{85}Kr sources, the ^{85}Kr gas can be released to the environment in a controlled manner.

FIG. 3. Example strategy for RLC management: advanced infrastructure.

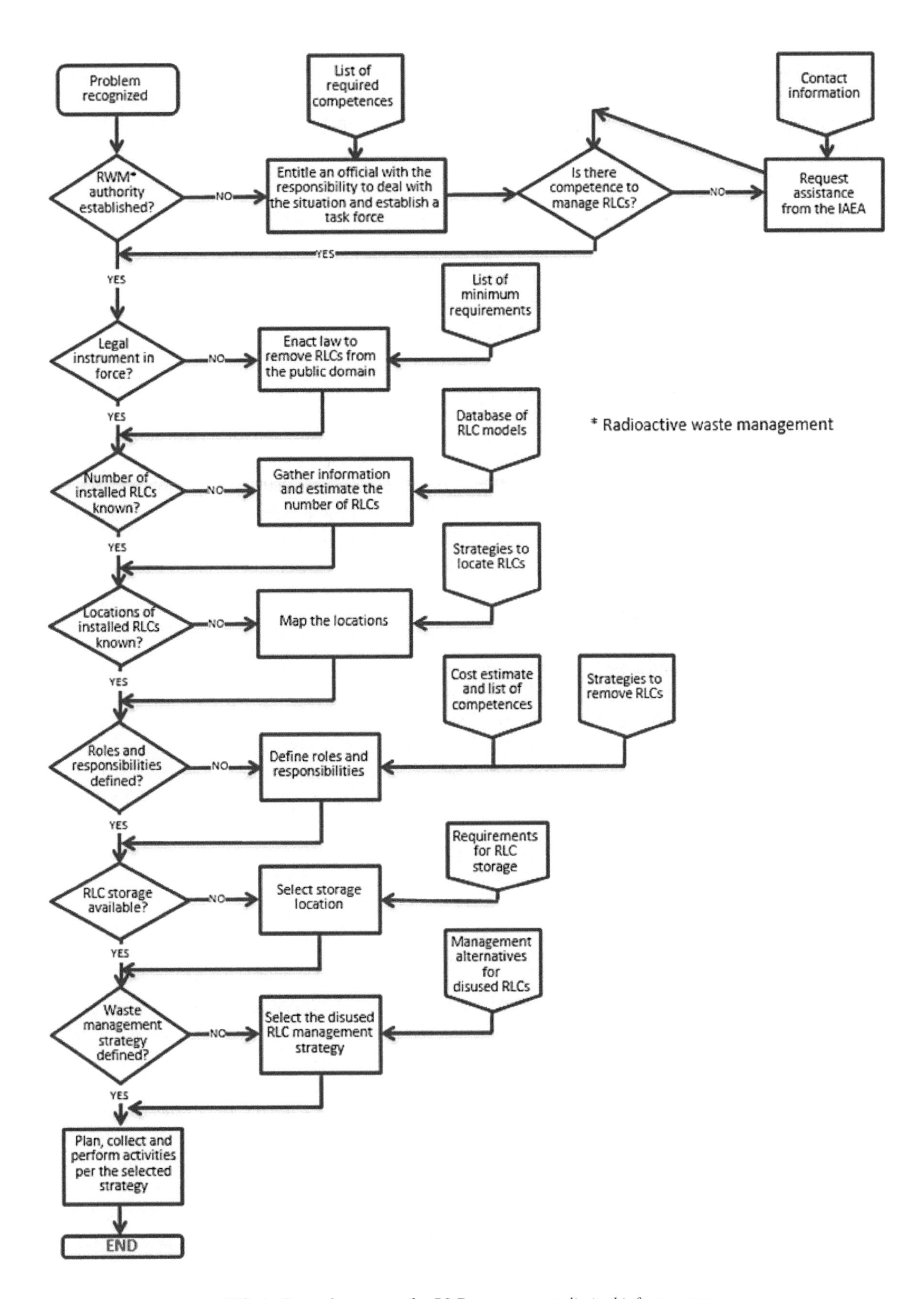

FIG. 4. Example strategy for RLC management: limited infrastructure.

management of RLC sources. The regulatory authority has to be involved in all activities, for example the approval of the work plan and the initial contact of the property owner. It is also assumed that long term management of sources will be conducted by licensed radioactive management organizations, which will be responsible for conditioning, storage and disposal.

4.4. TRAINING OF PERSONNEL

The government are to ensure that requirements are established for education, training, qualification and competence in protection and safety for all persons engaged in activities relevant to protection and safety [1]. The regulatory body will ensure the application of these requirements [1].

4.4.1. Radiation protection training

The removal/management personnel need to be trained on the basic radiation protection principles (e.g. time, distance, shielding) and be provided with the necessary individual monitoring, personal protective equipment (PPE) and radiation detection equipment to ensure that the operation is performed with a minimum of radiation exposure and in full compliance with the requirements of the established radiation protection legislation within the country. This training will include contamination control techniques and decontamination procedures, as it cannot be assumed that a source in an RLC is not leaking.

4.4.2. Occupational health and safety training

Workers need to be informed about the occupational risks involved and trained on necessary protection measures. Appropriate PPE will also be provided. Training on working at heights and the safe use of ladders and lifts will be required. The use of safety harness equipment will be required for work at heights. Safety considerations to minimize risk of electrocution while working close to power lines need to be taken into consideration. Training on the safe use of mechanical tools to detach the lightning conductor head from the pole has to be provided. Safety shoes and safety glasses also need to be worn.

4.5. FINANCING

Guidance from No. GSR Part 1 (Rev. 1) [29] states that "Appropriate financial provision shall be made for ... Management of radioactive waste, including its storage and disposal ... Management of disused radioactive sources".

Theoretically, this is also applicable to RLC management. However, since RLC management is essentially a legacy issue (they are no longer manufactured and/or their use is banned in some Member States), conventional financial provisions may not apply.

Therefore, Member States that have to remove and manage RLCs within their territory need to ensure that financial provisions are made. As mentioned previously, in principle, the 'owner pays' approach is preferred, but often that approach has to be replaced or subsidized by the governments involved. The senior management of organizations, in accordance with their accountabilities, are to ensure that provision is made for adequate resources and funding, including for the long term management and disposal of radioactive waste, IAEA Safety Standards Series No. GSR Part 2, Leadership and Management for Safety [33].

4.6. MANAGEMENT SYSTEM

According to Principle 3 of the No. SF-1 [28], "**Effective leadership and management for safety must be established and sustained in organizations concerned with, and facilities and activities that give rise to, radiation risks**". It is also stated that:

> "Safety has to be achieved and maintained by means of an effective management system. This system has to integrate all elements of management so that requirements for safety are established and applied coherently with other requirements, including those for human performance, quality and security, and so that safety is not compromised by other requirements or demands. The management system also has to ensure the promotion of a safety culture, the regular assessment of safety performance and the application of lessons learned from experience."

General requirements for a management system are established in IAEA Safety Standards Series No. GSR Part 2, Leadership and Management for Safety [34] and the recommendations in the IAEA Safety Standards Series No. GS-G-3.1, Application of the Management System for Facilities and Activities [35]. Guidance on the application of the management system for the processing, handling and storage of radioactive waste, including guidance on the application of the management system for the disposal of radioactive waste, is provided in IAEA Safety Standards Series No. GSG-16, Leadership, Management and Culture for Safety in Radioactive Waste Management [36].

The management system for the national RLC management programme has to cover all of the technical, safety, security, human resource and financial elements of the programme that are discussed in Sections 4–9 of this publication in the form of written work processes. A work process may cover one or more of the RLC management steps (e.g. one process for initial inventory verification, one for on-site work planning, one for source removal from RLC heads, one for immobilization with cement, etc.) More details on work process development and implementation are provided in No. GSG-16 [36].

The senior manager of the RLC programme will assign the responsibility for collecting and maintaining all work process documents and records to a suitably qualified individual. This individual has to report periodically to the senior programme manager about the status of the work processes. A typical but not exhaustive set of documents covering responsibilities, technical procedures, staff training, material documentation, safety and security measures, reporting and record keeping relevant to RLC work processes is as follows:

(a) A set of facility and operation licences issued by the regulatory authority.
(b) A set of written procedures for all technical operations (initial RLC inventory and verification, plans for site operation for RLC removal, packaging for transportation, transportation, treatment/conditioning, storage).
(c) A list of organizations and individuals responsible for the above listed operations, including their qualifications, licences and training documentations. External contractors (e.g. for removal of RLCs from the original locations) will also have to submit these data.
(d) A list of the records and their contents that have to be prepared during all of the above listed operations and submitted to the relevant databases (e.g. the Information Database — IDB).
(e) A list of quality requirements for all used material (e.g. cement and steel capsules used for conditioning of RLCs) and purchase documentation for all materials actually used.
(f) Written quality control procedures (e.g. for welding quality or quality of concrete used for immobilization) and records of all actual quality control results.
(g) Written radiation protection procedures, person(s) responsible for radiation monitoring (including monitoring of possible surface contamination) and records of monitoring results.
(h) Written security and emergency preparedness procedures.

The management system based on the set of responsibilities, written procedures, materials, operational and quality control records will be approved and supervised by the senior manager of the RLC programme.

5. PREPARATORY WORK (BEFORE RADIOACTIVE LIGHTNING CONDUCTOR REMOVAL)

A key part of managing RLCs in the public domain is to locate them and create an inventory describing, among other things, where they are located, physical properties, barriers to removing them, etc. (Section 5.1). Communicating the need to remove RLCs to the public can help locate them. See Appendix II for an example of a public awareness programme.

5.1. LOCATING RADIOACTIVE LIGHTNING CONDUCTOR

A national strategy for locating RLCs has to involve a search for devices in use or disused. This section provides guidance on methodologies for conducting such searches, which may be administrative and/or physical searches.

An administrative search may identify evidence of a lost source and lead to a physical search for the source. Administrative searches are also used to prioritize physical searches. A physical search will involve attempts to identify radioactive sources both visually and with the use of radiation detectors [37].

5.1.1. Administrative searches

Establishing a countrywide inventory begins with conducting administrative searches of known historical records and conducting searches to locate as yet unknown records. Information is gathered without the use of radiation detection equipment or visual searches.

Two key aspects of administrative searches are determining the most useful source of information and determining the best method for collecting the information from that source. The initial searches will cover all lightning conductors, not just RLCs, since some may be mistakenly identified as not being RLCs when, in fact, they are RLCs.

The person or institution where the desired information might currently reside has been termed the 'target'. One of the first tasks of an administrative search is the identification of the potentially useful targets for searches. This is useful for identifying targets for communication purposes. Such targets may include:

(a) Government authorities (federal, regional* (state/provincial) and municipal*)
(* = likely the best targets based on experiences in Belgium and France);
(b) Non-government or international organizations;
(c) Users and owners;
(d) Manufacturers, distributors and installers;
(e) Individual workers;
(f) The general public.

The tools or methods used to gather data during an administrative search can be broadly grouped into three types, namely: broadcast media, records searches and interviews. The appropriate tool to use in each circumstance will be dependent on the reason for, and the extent of, the search. More details on both targets and tools are given in IAEA Safety Standards Series No. SSG-19, National Strategy for Regaining Control over Orphan Sources and Improving Control over Vulnerable Sources [37].

5.1.2. Physical searches

A physical search primarily involves the development of a search plan, after which a search team of one or more persons goes to physically locate RLCs, both visually and using radiation detectors. Generally, a physical search is conducted following an administrative search. However, a search programme is an iterative process and, in certain circumstances, a physical search may start at the same time as, or even before, an administrative search. Because search teams may encounter radioactive sources, the need for radiation protection measures for these individuals has to be considered.

Establishing a list of possible locations where RLCs can be found has to involve the regulatory authority. The regulator may have access to records indicating the possible location of RLCs. If no records can be found, using public advertisements with radio, television or newspapers showing pictures of lightning conductor devices and requesting notification if they have seen one of these devices may be another technique to locate RLCs within the country (see the example in Appendix II). If legislation exists in the country to support this effort, it needs to be referenced during the campaign. If it is known who the manufacturer/distributor/installer was or the type of lightning conductor that was installed, contacting one of them may help in locating records for the locations where the units were installed. Specific mailings to the owners of facilities such as high buildings, hotels, churches, large apartment buildings and public buildings may also locate missing lightning conductor sources. The initial minimum information needed on each RLC is the location and the property owner.

5.2. INITIAL INVENTORY VERIFICATION

Before planning to remove an RLC, inventory verification has to be performed. This involves a site visit by a person or persons delegated by the organization responsible for RLC management. The purpose of site visits is to determine the condition of RLCs and to plan for their removal or to improve their condition and/or security if an exemption is granted to allow the continued used of a specific one.

Before any site visit, the property owner ought to be contacted and informed of the plans to investigate the condition of the RLC. The plan to eventually remove the source from the RLC, and possibly the lightning conductor itself, needs to be made clear to the property owner. An RLC sources fact sheet will be given to the property owner. A sample fact sheet is provided in Appendix II.

During the site visit, a visual confirmation of the lightning conductor is made and a photograph ought to be taken to record its current location and status. A radiation survey has to be carried out to determine if a radioactive source is present or if the site is contaminated. An appropriate survey meter will be used to take measurements at different distances from the RLC head if possible. Depending on the survey methodology and the design of the RLC, it may be possible to deduce the radionuclide(s) present.

Appendix I includes images of alpha, beta and gamma type RLCs, as well as images of conventional (non-radioactive) lightning conductors for comparison purposes.

5.3. PREPARATION OF SITE WORK PLAN(S)

A written work plan will be a part of the authorization of the operator to perform the work. The work plan needs to include:

(a) An outline of the work to be performed, which includes removal, transport and storage;
(b) A written, up to date risk assessment of the work to be performed;

(c) Written procedures incorporating radiation, occupational health and environmental safety considerations at each step;
(d) Specification of the requirements for skills, training and record keeping;
(e) Allocation of responsibilities, for example for radiation protection.

Key safety considerations ought to include not only the fact that a radioactive source is being handled but also the fact that typically this work includes other occupational hazards. As examples, work may require the use of long ladders or lifts to work at heights, electrical power lines may be in the way, and access to worksite may be difficult. Work will never proceed unless it can be ensured that it can be done safely. Equipment is needed to access and remove the lightning conductor head, test and decontaminate the surrounding area if needed, and shield the source and package it for transportation to the conditioning facility. Of note, conventional hazards (like falling from height) are likely greater than radiological hazards.

Radiation detection equipment for individual monitoring of operators is required. Personal protective equipment needs to be provided. The area around the source will need to be monitored to determine if contamination is present [38] and, if so, the ability to decontaminate the area will be required. Typical decontamination methods include wiping, scrubbing, flushing and soaking. The rule of thumb is to start with a method that creates the least amount of secondary radioactive waste, and so it is worthwhile to try to locate the contamination with the use of survey instrument and target the area(s) to be cleaned. Secondary radioactive waste will have to be packaged and transported to a licensed radioactive waste management facility.

The type of approved packaging for transport to the storage facility needs to be defined. The transportation vehicle and driver have to meet all radioactive material transportation requirements for the country.

5.4. RECORD KEEPING

Prior to removal of the RLC from its installed location, a data file has to be created. The file needs to be structured to readily record data collected throughout the RLC removal operation and subsequent management for each RLC. Figure 5 depicts the various stages of record keeping, from locating RLCs up to disposal of their radioactive components. All information has to be retained in duplicate at separate locations. Planning needs to include plans for the migration of data to future information systems as technology changes.

During the notification and site visit step, including the administrative process of transferring the RLC to regulatory control, all information, to the extent possible, needs to be collected and recorded in an information database (IDB) in order to ensure the traceability of sources and the associated radioactive waste. It is up to the regulatory authority to assign responsibility for creating and maintaining the IDB. The organization implementing the long term management of RLCs (see Section 4.3) may be the most appropriate holder of the IDB. All other parties involved in the different activities (location of RLCs, initial verification, removal, transportation, treatment/conditioning, storage) will provide the necessary data. The regulatory authority will keep a copy of the IDB, which may not necessarily contain all technical details that are needed to implement the different activities.

The first set of IDB data will include:

(a) Reference ID(*)[4]: assigned by IDB holder, ought to be as simple as practicable, used to link the various data to be collected.
(b) Location and accessibility: used to plan resources and operations. The data to be recorded are:
    (i) Address(*): city, street, coordinates, etc.;
    (ii) Location(*): roof, cornice, etc., useful for planning the tools needed;

[4] * = minimum requirement.

  (iii) Height of the conductor: useful for planning the tools needed;
  (iv) Accessibility(*): window, trap door, staircase, etc.;
  (v) Photographs;
  (vi) Status of the RLC(*): intact, fallen, damaged, corroded, etc.;
  (vii) Surface contamination(*): useful for planning decontamination;
  (viii) Dose rate(*): of the RLC head at specified distances (e.g. 50 cm) from and at selected points in the surrounding area.

(c) Owner and owner's declaration(*): identification of the owner (company or official body, if possible) and the owner's name or its legal representative that signed the agreement to remove the RLC and transfer responsibility to the regulatory authority or another body that acts on behalf of this authority, depending on the laws or norms in each case.

(d) Manufacturing data: helpful to identify the nominal physical and radiological characteristics of the head to be removed and to cross-check its possible logging in the manufacturer's database. The information to be recorded is:
  (i) Manufacturing company;
  (ii) Model;
  (iii) Serial number;
  (iv) Date of delivery/date of manufacture;
  (v) Radionuclide: by certificate, measurement or suspected radionuclide by dose rate;
  (vi) Activity: nominal or measured;
  (vii) Copy of invoice or certificate from manufacturing company.

Separate data fields for each parameter are suggested since much of the information cited may not be available.

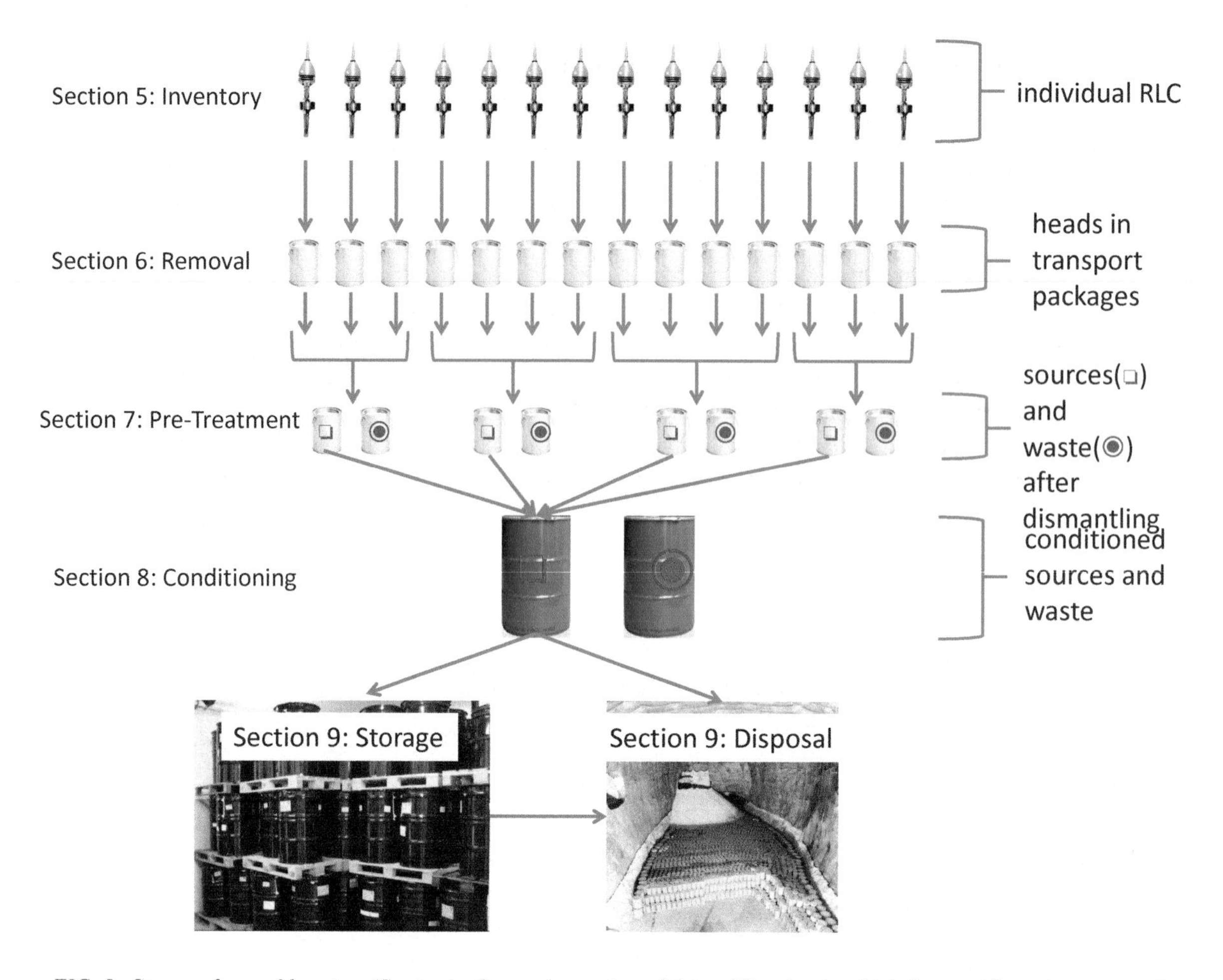

FIG. 5. Stages of record keeping. 'Section' refers to the section of this publication in which the specific topic is covered.

6. SITE WORK: RADIOACTIVE LIGHTNING CONDUCTOR REMOVAL, HEAD REMOVAL, PACKAGING AND TRANSPORT

This section describes the removal of RLCs from the public domain. This includes the removal of RLCs from buildings and structures, packaging for transport and storage of their heads in an unconditioned state in an appropriate storage facility. It is assumed that the minimum infrastructure to support these activities, covering staff training, radiation protection, occupational health and safety, contamination control, decontamination, characterization, etc., is available.

This section provides an overview of what qualified, trained and authorized persons carry out and is not meant to be guidance for untrained members of the general public. It is more of a cautionary measure to inform the public of the hazards involved in RLC removal.

Once an RLC head is taken to a storage facility, it may remain there until additional heads are collected or until options are available for further management (Section 9.1).

6.1. REMOVAL OF RADIOACTIVE LIGHTNING CONDUCTOR

6.1.1. Preparing for removal arrangements

Removing RLCs that are at the same or adjacent locations will allow for the most efficient use of time and equipment. Making arrangements in advance with each property owner so they know what to expect prior to arrival can help ensure access to the property. The verified inventory information ought to be studied in advance to determine the type of tools needed to access the lightning conductor head. The radioactive source is securely installed in the head of the lightning conductor in all models, hence the whole head will be removed from the pole and the actual radioactive source will be removed later at a conditioning facility.

Before starting any activities at a location where an RLC will be removed, the area has to be roped off in order to keep the public at a proper distance. Either a tall ladder or a mechanical lift will be required in most cases to remove the lightning conductor head, unless the entire pole has fallen or been taken down. Arriving at the site with the proper equipment to gain access to the RLC head is critical to the job going smoothly. Long ladders or mechanical lifts and assistance with the lifts may be available from the local fire department. The regulatory body may be of assistance in soliciting the help of the local fire department and use of their lifting equipment for this purpose. All use of ladders and lifts ought to be done with safety considered first. Lifts could be preferred over ladders to allow for the operator to have his/her hands free to use tools to remove the RLC head. A lift also is preferable for gamma sources because it also allows the operator to lift additional shielding material to the source head, if required. Great caution in all work around electrical and power lines is vital.

There are many different types of RLC supporting structures. Typically, they are either roof mounted poles that are ~6 m high or free standing poles at heights of 17–21 m, usually secured with anchor wires. Cutting of the anchor wires may also be required.

6.1.2. Head removal

A trained, designated operator will mount the lift with the necessary tools to remove the head from the pole. If it is difficult to remove the head, it may be necessary to cut the pole to remove the entire head. If the pole is contaminated, the whole pole may have to be removed.

6.1.2.1. Alpha emitting source head removal

Alpha emitting sources (^{241}Am and ^{226}Ra) have a history of loose and dispersible contamination problems within RLC heads. Contamination may also be found in the surrounding area close to the head, such as on roofing material directly under the source head. Contamination problems have not usually been found with gamma emitting sources ($^{154/152}$Eu and ^{60}Co), but it is not impossible, so all lightning conductor heads ought to be treated as potentially leaking loose contamination until proven otherwise. The operator needs to wear protective clothing and place the entire head in a plastic bag to contain any possible contamination until confirmation is established that it is contamination free.

Care needs to be taken while disturbing roof areas directly under the lightning conductor, especially in the case of the alpha emitting sources, so as not to spread any possible contamination.

6.1.2.2. Beta/gamma emitting source head removal

Beta/gamma sources are less likely to have contamination issues than alpha sources. Gamma emitting sources ($^{154/152}$Eu and ^{60}Co) may have high levels of activity and dose rates that require operators to work at a distance from the RLC or may require additional radiation shielding.

Based on the amount of time it will take to remove the head from the pole by unscrewing it (first choice) or cutting (second choice if unable to unscrew), it has to be determined if shielding needs to be added to the unit to decrease radiation dose to the operator. Handling a heavy lead shield while on the lift and trying to remove the head from the pole may also result in awkward handling and more of a danger than simply utilizing time and distance to minimize radiation exposure. Each situation needs to be assessed independently. If the operator is able to keep back from the source head as far as possible while removing the head and then quickly place the head into a shielded box that is also on the lift, this may also be an acceptable practice.

Operators have to wear personal dose monitoring devices and a radiation survey meter when removing RLCs. Pictures of RLC heads can be found in Appendix I and the annexes.

6.2. PACKAGING AND TRANSPORT OF RADIOACTIVE LIGHTNING CONDUCTOR HEADS

Transporting an RLC head with its radioactive source to the waste storage or conditioning facility requires that the unit be packaged per applicable national regulations for transporting radioactive material on public roads. If no national regulations for transporting radioactive materials exist, the packaging and transportation of the source have to meet the requirements of the IAEA Safety Standards Series No. SSR-6 (Rev.1), Regulations for the Safe Transport of Radioactive Material [39].

It is assumed that the residual activity of RLC sources will allow them to be transported in a Type A package (see Table 1).

Secondary waste, such as contaminated components, contaminated PPE and decontamination waste, for example towels and wipes, will be at a much lower activity level and likely can also be transported in Type A packaging. Each drum or container will have at least one radiation warning symbol (trefoil symbol) attached and the package ID (see Section 6.3).

6.3. RECORD KEEPING

Information collected during the removal of RLCs and transport to the next step in the management scheme is summarized below (the second set of IDB data). All information has to be retained in duplicate at separate locations.

TABLE 1. EXAMPLES OF TYPE A PACKAGE LIMITS

Radionuclide	Activity limit (material not in special form)
Eu-152	1 TBq (27 Ci)
Eu-154	600 GBq (16.2 Ci)
Co-60	400 GBq (10.8 Ci)
Am-241	1 GBq (27 mCi)
Ra-226	3 GBq (81 mCi)

6.3.1. Packages with heads and sources

(a) Removal package ID (on outside of container) (see Fig. 6);
(b) Reference ID (see Section 5.4);
(c) Removal date;
(d) Operator information (e.g. company name, contact details);
(e) Description of package contents (e.g. head with 10 cm of pole attached);
(f) Radionuclide(s);
(g) Activity (activities);
(h) Dose rate contact: maximum rate on contact;
(i) Dose rate @ 1 m: rate at 1 m from the container;
(j) Surface contamination;
(k) Mass;
(l) Photographs (container contents, container).

6.3.2. Packages with secondary waste

(a) Removal package ID (on outside of container);
(b) Reference ID (see Section 5.4);
(c) Removal date;
(d) Operator information (e.g. company name, contact details);
(e) Description of package contents (e.g. PPE, wipes, contaminated parts);
(f) Radionuclide(s);
(g) Activity (activities);
(h) Dose rate contact: maximum rate on contact;
(i) Dose rate @ 1 m: rate at 1 m from the container;
(j) Surface contamination;
(k) Mass;
(l) Photographs (container contents, container).

FIG. 6. Example of a removal package ID. Courtesy of the Centre for Radiation Protection and Hygiene, Cuba.

7. MANAGEMENT OF THE RADIOACTIVE LIGHTNING CONDUCTOR HEADS AT CONDITIONING AND/OR STORAGE SITES

7.1. ADMINISTRATIVE ISSUES

The first action taken is to record the package removal ID (see Section 6.3.1) within the IDB when packages with heads and sources are received at the conditioning and/or storage facility. The next action is to decide which management option(s) to take, depending on the sources, radionuclides and available technologies and facilities, such as:

(a) Store packages as received without further action (may require relabelling);
(b) Consolidate heads with sources and repackage;
(c) Remove sources from heads and package sources and other waste separately;
(d) Treat/condition sources and other wastes (e.g. encapsulation);
(e) Store newly created packages.

Radioactive source conditioning has to be carried out by qualified personnel. The source conditioning procedure will include detailed consideration of the following:

(a) Regulatory and personnel requirements;
(b) Records;
(c) Inventory per package;
(d) Long term management (store, dispose, transport to storage and disposal).

ALARA principles have to be applied to all the handling operations to minimize radiation exposure. As many of the RLC sources were manufactured several decades ago, degradation leading to contamination could be a possibility, with the foil sources being most at risk from damage by the elements. Therefore, in handling such sources, regular contamination checks and contamination control ought to be carried out. Special precautions to contain and reduce the exposure to personnel and the environment from the spread of contamination will be required for damaged and leaking sources. Personal dosimeters, area monitors and surface wipes for loose contamination, relevant to the type of radiation and contamination being handled, will be used.

7.1.1. Regulatory issues

It is the responsibility of the operator of conditioning and storage facilities to meet technical and functional specifications in accordance with legislative requirements. Specific guidance on regulatory requirements for the design and operation of waste processing (including conditioning) and storage facilities is given in IAEA Safety Standards Series No. SSG-40, Predisposal Management of Radioactive Waste from Nuclear Power Plants and Research Reactors [40], IAEA Safety Standards Series No. SSG-45, Predisposal Management of Radioactive Waste from the Use of Radioactive Material in Medicine, Industry, Agriculture, Research and Education [41] and IAEA Safety Standards Series No. WS-G-6.1, Storage of Radioactive Waste [42]. The design of facilities and all operations carried out within them have to be authorized (licensed) and inspected, with conditions of authorization enforced by the competent authority. In the case of a national programme for RLC management, treatment/conditioning of RLCs may be carried out in an existing facility or in new ones.

When the treatment/conditioning and storage of RLCs will be carried out in a pre-existing facility licensed for operations other than RLC conditioning and storage, the operator has to investigate whether the facility and planned operations with RLCs meet the requirements of the existing licence. Even if the existing licence allows for such operations, the operator will notify the regulator about the new operations. If the operations with RLCs do not meet the requirements of the actual licence, the operator will have to prepare a request for amendment of the existing licence and submit it to the competent authority. When treatment/conditioning and storage of RLCs will be carried out in a new facility, the operator will have to prepare a new licence application and submit it to the competent authority.

7.1.2. Personnel requirements

The conditioning process requires skilled personnel with an appropriate level of skills, knowledge and practical experience for high quality work. The number of persons needed is not prescribed. Nevertheless, at least two persons ought to be present during any operation with radioactive material.

7.1.3. information about the sources in radioactive lightning conductor heads

Before detailed planning is possible, it is necessary to have all relevant facts about the sources to be conditioned, such as:

(a) Source type, geometry and dimension;
(b) Physical and chemical form;
(c) Activity;
(d) Name and address of the last user of the source;
(e) The measured dose rate at a known distance.

This information might already be in the IDB (see Sections 5.4 and 6.3). Other possible sources of this information may be with the manufacturer of the source or the national regulatory authority.

7.1.4. Conditioning and inventory per package

Conditioning of radioactive sources in a cement lined 200 L drum or in another suitable container will result in a package that is suitable for both transportation and storage. Where a preferred disposal option has been identified (e.g. borehole disposal, geological disposal), then radioactive sources are best conditioned in a manner that is consistent with that disposal option. A 200 L drum containing conditioned sources that meets Type A package requirements would be the easiest to transport [39]. Example Type A package limits are listed in Table 1. The surface dose rate of the package cannot exceed 2 mSv/h on contact and 0.1 mSv/h at 1 m.

However, the limits cited above may not necessarily be suitable for storage. Depending on the licence for the storage facility, higher values may be accepted for storage. Without the intention to dispose of the sources in the near future, advantage could be taken of the half-life of ^{60}Co and $^{152/154}$Eu and higher activity levels of such radionuclides may be loaded in a drum. For example, during a storage period of 20 years, the activity is reduced by a factor of 2.8 for ^{152}Eu and by a factor of 14 for ^{60}Co. However, higher limits may preclude Type A package transport, resulting in increased costs associated with packaging and transport.

After conditioning, a unique identifier will be placed on the outer surface of storage containers for both sources and secondary waste and recorded in the IDB along with package location. A temporary label will be affixed until the contents have been finalized. Once packaging is complete, the containers will be permanently labelled (e.g. by engraving.). Labels need to be readable for decades without reliance on any specific technology. See Section 7.4 regarding record keeping.

Each drum or container has to have at least one radiation warning symbol (trefoil symbol) attached. It may be necessary to ensure that sources are retrievable from packages (e.g. not fixed within a cement matrix) depending on the disposal option.

7.2. INITIAL ACTIONS WHEN RADIOACTIVE LIGHTNING CONDUCTOR HEADS ARE RECEIVED

The packages of RLC heads containing their sources that arrive at the conditioning and/or storage facility need to be stored safely and securely, and their package IDs need to be recorded in the IDB. If management option 1 is chosen (see Section 7.1), no further actions are taken.

If heads and sources are to be removed from their containers, the facility ought to have shielded areas in which to handle them. Due to possible high dose rates from some gamma sources, remote handling tools ought to be used wherever possible, and a radiation survey meter always has to be available.

Sources can be left in their heads or removed — Section 7.3 gives an overview of the steps needed if sources are removed. Leaving sources in their heads could be based on economic (there are only a few RLCs to manage) or technical reasons (handling equipment is not readily available).

7.3. TREATMENT/CONDITIONING OF SOURCES REMOVED FROM HEADS

Figure 7 illustrates some of the possible steps involved in separating sources from their heads and the subsequent management of the sources.

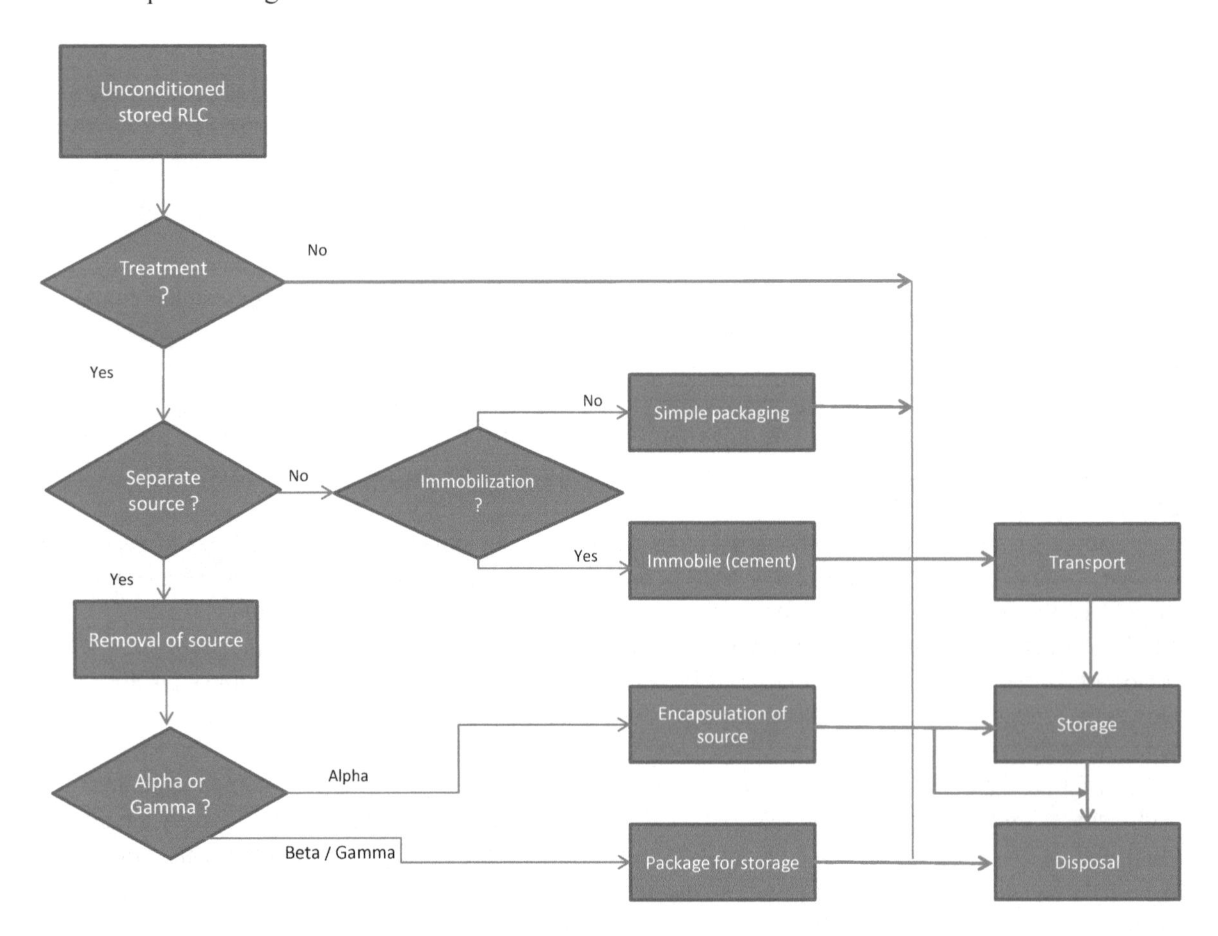

FIG. 7. Treatment and conditioning options for RLC heads.

Immobilization may preclude retrievability and only ought to be considered if there is no intention to retrieve sources from packages. To the extent practicable, radioactive waste ought to be stored according to the recommendations for passive safety [42].

The separate treatment/conditioning of RLC containing radionuclides with half-lives above and below 30 years may be considered, which will facilitate later planning and preparation for disposal.

7.3.1. Alpha emitting sources

Contamination control is the principal concern when handling an alpha foil from an RLC (foil or wire with ^{241}Am, ^{226}Ra or other alpha source). There may be loose contamination from ruptured, oxidized or otherwise missing foils. The handling of alpha foils and other disused sources could lead to the ingestion of contamination. Conventional, industrial and radiological protection safety rules and procedures need to be observed. Adequate ventilation and filtration systems need to be in place to protect personnel and the environment from airborne contaminants if loose contamination is expected. Appropriate personal protective equipment is also important to prevent internal and external exposure to contamination.

Alpha emitting foils have to be handled or conditioned in a filtered, ventilated glovebox, or where this is not viable, a provisional working arrangement may be sufficient (see Fig. 8).

In the case of alpha emitting sources, the contaminated RLC head, the removed source foils and the resulting wastes will be double bagged in heavy duty polyethylene bags before being packaged into 200 L (55 gallon) drums for storage.

7.3.2. Beta/gamma emitting sources

If no contamination is found, sources can likely be removed from the head without being in a glovebox or bag. A screwdriver or other similar device with a hammer is typically used to pry off the top metal enclosure cap from the head of the RLC to expose the lead shield. Inside the lead shield is the source. Remote tools need to be used to lift the lead shield containing the source out and place it into

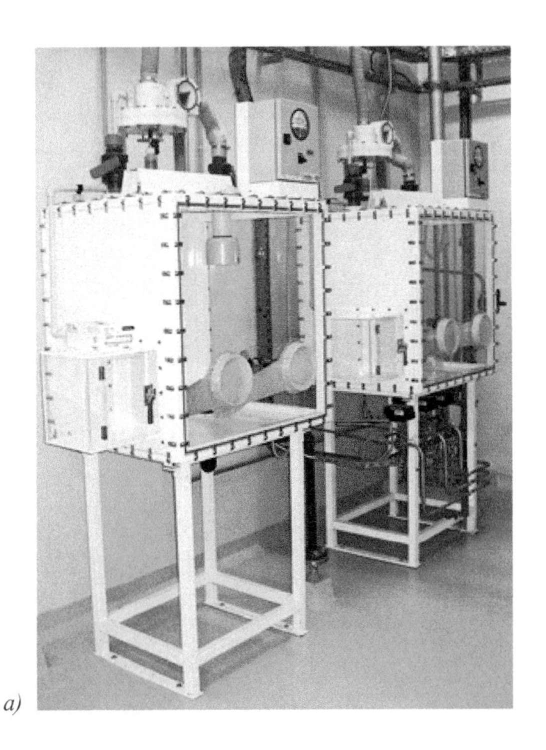

a)

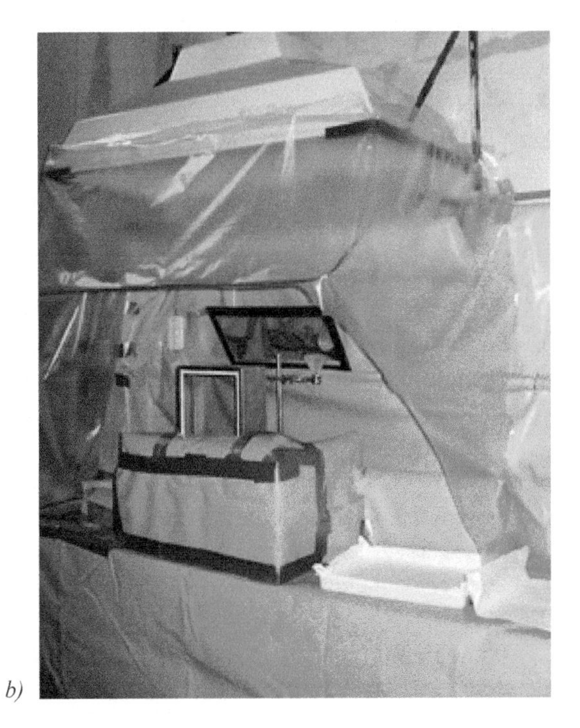

b)

FIG. 8. Examples of (a) a glovebox and (b) a provisional working arrangement for handling alpha sources.

a larger lead shield for eventual conditioning into a storage drum. Beta/gamma sources are not to be manipulated by hand with or without gloves.

7.4. RECORD KEEPING

Records ought to be created and managed to preserve the results of the waste management activities; these may be needed in the future to implement both the expected waste management option and other options that may have to be implemented [36].

The information collected during treatment of RLCs and their sources and their transport to the next step in the management scheme is summarized below (the third set of IDB data). All information has to be retained in duplicate at separate locations. If sources/waste are conditioned, then additional record keeping requirements will apply (see Section 8).

As noted in Section 7.1, a unique identifier will be placed on the outer surface of storage containers for both sources and secondary waste and recorded in the IDB along with package location. A temporary label will be affixed until the contents have been finalized. Once packaging is complete, the containers will be labelled permanently (e.g. by engraving.). Labels ought to be readable for decades without reliance on any specific technology.

7.4.1. Packages with heads and sources or just sources (treated or untreated)

(a) Waste package ID;
(b) Reference ID for each RLC processed (see Section 5.4);
(c) Packaging date (date package closed);
(d) Operator information (e.g. company name, contact details);
(e) Description of package contents (e.g. sources only);
(f) Radionuclide(s);
(g) Activity (activities);
(h) Dose rate contact: maximum rate on contact;
(i) Dose rate @ 1 m: rate at 1 m from the container;
(j) Surface contamination;
(k) Mass;
(l) Photographs (container loading and contents, container).

7.4.2. Packages with secondary waste (treated or untreated)

(a) Waste package ID;
(b) Reference ID for each RLC processed (see Section 5.4);
(c) Packaging date (date package closed);
(d) Operator information (e.g. company name, contact details);
(e) Description of package contents (e.g. PPE, wipes, contaminated parts);
(f) Radionuclide(s);
(g) Activity (activities);
(h) Dose rate contact: maximum rate on contact;
(i) Dose rate @ 1 m: rate at 1 m from the container;
(j) Surface contamination;
(k) Mass;
(l) Photographs (container loading and contents, container).

8. CONDITIONING OF THE SOURCES AND SECONDARY WASTE

8.1. SELECTION AND QUALIFICATION OF THE CONDITIONING METHOD

Conditioning, according to the IAEA Safety Glossary [43] is "Those *operations* that produce a *waste package* suitable for handling, *transport*, *storage* and/or *disposal*."

Conditioning of DSRSs ensures containment of the radioactive material, provides for greater confinement of leaking sealed sources, provides sufficient radiation shielding, reduces storage/disposal volume by allowing for consolidation of multiple sources into a single storage/disposal container, facilitates transport operations and contributes to safety and security [4].

Conditioning needs to take into account the fact that acceptance criteria for future management of radioactive sources may not yet be specified. Conditioning techniques that could cause difficulties at a later stage (e.g. need for difficult and costly reconditioning) ought to be avoided. It is preferable to avoid all irreversible conditioning of radioactive sources, particularly direct embedding in a matrix, which might not necessarily be compatible with later management stages. If a waste package includes lead for shielding, it is best to not embed DSRSs in a cement matrix in the package. This will allow easy removal of DSRSs from the package to separate them from the lead shielding before emplacement in a disposal facility.

Conditioning techniques need to allow for easy retrieval of sources at a later time. It is advisable to form a well inside the concrete by using, for example, a small steel bucket. The conditioned sources are placed in the well. A steel bar welded across the well deters unwanted removal of sources. A lead shield on top of the well provides radiation shielding. Drums shielded with lead and concrete can weigh more than 200 kg, and this will enhance the security of the drum because the drum cannot be moved easily.

Because they have different characteristics, alpha emitting sources (^{241}Am or ^{226}Ra) and gamma emitting sources ($^{152/154}$Eu or ^{60}Co) need to be conditioned separately. For gamma emitting sources with different half-lives, such as $^{152/154}$Eu or ^{60}Co, the same conditioning method can be applied, but they need to be conditioned and stored in different drums.

Taking into account these constraints for the conditioning, the storage has to be safe, especially with regard to radiation, contamination, fire risks and physical safety.

Consideration ought to be given to the length of time, the extremes of temperature and the humidity that the waste drums will be exposed to for the duration of storage. The durability of the package, including identification marks, needs to be in accordance with the length of the foreseen storage period (see Section 7.1). Only approved drums from a qualified supplier will be used. Stainless steel and coated carbon steel drums maintain their integrity for a long period of time.

8.2. EQUIPMENT AND MATERIAL REQUIRED

When the details of the conditioning process are decided, all equipment and material required will be collected. Although there might be some differences depending on exactly how the conditioning is to be carried out, the following may typically be required.

General:

(a) Personal dosimeter;
(b) Calibrated dose rate meters and contamination monitors;
(c) Labels (for storage and transportation);
(d) A forklift truck.

Shielded drum or other suitable container:

(a) Two hundred litre painted or coated carbon steel or stainless steel drums;
(b) Additional shielding material to meet the dose rate criteria on the package, usually lead;
(c) Mould (e.g. bucket) for preparing a cavity inside the 200 l drum;
(d) Cement, sand and water according to specified mixture;
(e) Cement mixer;
(f) Vibrational or manual compaction of mortar;
(g) Steel reinforcement bars.

8.3. CONDITIONING OF ALPHA SOURCES

Conditioning provides high integrity containment for the sources and minimizes radiation dose rates for storage and transportation. Encapsulation of alpha emitting sources is required for contamination control. If multiple sources are in a steel bucket, the lid of the bucket will be welded and tested for tightness. The number of sources in a bucket depends on their activities and the size of the bucket. Steel bars will be welded across the opening of the cement liner and drums will have a gross weight of at least 200 kg.

The method described is primarily used to add weight to the drum to make it difficult to move for security reasons and incidentally provides shielding. The package has to be designed to withstand storage periods of several decades.

The method of source encapsulation for alpha emitting sources is detailed further in Ref. [4].

8.4. CONDITIONING OF BETA/GAMMA SOURCES

Although contamination control is typically not an issue, as with beta/gamma sources, further encapsulation is still suggested. Shielding of such sources is sometimes required in order to reduce the radiation levels at the surface of the package to acceptable levels and to provide physical security for the sources. The design of a shielding package has to take into account the following factors:

(a) Total radioactivity;
(b) Retrievability;
(c) Physical security;
(d) Radiation protection;
(e) The storage period.

If it is intended that the shielded package will be used for transportation of the sources in the near future, the total radioactivity contained within the drum needs to be limited to Type A package values (see Section 6.2). During storage it is advisable to keep the conditioned sources retrievable, to avoid prejudicing their further management. Steel bars will be welded across the opening of the cement liner. Packages ought to withstand storage periods of several decades.

It is suggested that the shielded container consist of a lead liner or a lead container, contained in a concrete and steel overpack. The lead liner or container needs to be of appropriate wall thickness and of sufficient diameter to accept the sources and they ought to be packed in a manner that allows for their retrieval at a later date. The use of a steel container inside the lead shielding can facilitate retrieval. It needs to be noted that drums filled with cement and lead in this manner may weigh up to ~400 kg and may require a special lift to move.

8.5. MARKING AND LABELLING OF PACKAGES

Each container has to have a record of the contents within and at a minimum this will include:

(a) The type and quantity of sources contained (e.g. six foils, two pellets);
(b) Radionuclide(s);
(c) Total activity content in units of Bq or Ci;
(d) Surface dose rate specified in units of Sv/h or R/h;
(e) Date drum was sealed.

Each drum will be identified by a conditioned waste package ID, with records maintained in the facility and copies provided to the regulatory authority.

The information on the drums or containers will be engraved or stamped on aluminium or stainless steel labels and placed in suitable positions both within and secured to the outside the drum.

Each drum or container will have at least one radiation warning symbol (trefoil symbol) attached to the drum.

8.6. SAFETY AND SECURITY CONSIDERATIONS

During the conditioning process precautions need to be taken to restrict radiation exposures to a minimum according to ALARA principles. Work involving the handling of radioactive sources has to be carried out in a controlled area. Contamination and radiation monitoring need to be carried out during and after all work. Since RLC heads or sources may be contaminated, work in a glovebox is suggested. Prior to transportation of the sources, radiation and contamination levels have to be within the limits given in the No. SSR-6 (Rev.1) [39].

All persons engaged in the work have to be provided with personal dosimeters that are evaluated periodically (e.g. monthly for long campaigns) or at the end of a campaign. Consideration ought to be given to the requirement for finger ring dosimeters and personal alarm dosimeters. During the production of a shielded package, all safety precautions need to be observed, such as wearing protective clothing, safety glasses and steel toed shoes. Handling of heavy packages and lead have to be given special safety consideration.

In order to avoid misuse of such sources, security measures during all operations (including storage), as described previously, also need to be observed. Relevant information may be found in Ref. [31].

Further guidance on the safety and security of radioactive sources is given in the Code of Conduct on the Safety and Security of Radioactive Sources [44] and in IAEA Safety Standards Series No. RS-G-1.10, Safety of Radiation Generators and Sealed Radioactive Sources [45].

8.7. MANAGEMENT OF SECONDARY WASTE

Contamination may occur during all stages of RLC operation. This is especially valid for alpha sources.

However, only dry or damp material may be expected to be contaminated by either leaking sources or decontamination operations.

Such material will be collected, placed and tied off in plastic bags and stored in proper steel drums. Treatment or conditioning in the future may be considered. Disposal in a licensed facility without further action may also be an option.

8.8. RECORD KEEPING

The information collected during the conditioning of RLCs, their sources and secondary wastes, and their transport to the next step in the management scheme, is summarized below (the fourth set of IDB data). All information has to be retained in duplicate at separate locations.

8.8.1. Packages with heads and sources or just sources

(a) Conditioned waste package ID;
(b) Reference ID for each RLC processed (see Section 5.4);
(c) Conditioning dates for each RLC processed;
(d) Packaging date (date package closed);
(e) Operator information (e.g. company name, contact details);
(f) Description of package contents (e.g. sources only);
(g) Conditioning method(s) used;
(h) Radionuclide(s);
(i) Activity (activities);
(j) Dose rate contact: maximum rate on contact;
(k) Dose rate @ 1 m: rate at 1 m from the container;
(l) Surface contamination;
(m) Mass;
(n) Photographs (container loading and contents, container).

8.8.2. Packages with secondary waste

(a) Conditioned waste package ID;
(b) Reference ID for each RLC processed (see Section 5.4);
(c) Conditioning dates for each RLC processed;
(d) Packaging date (date package closed);
(e) Operator information (e.g. company name, contact details);
(f) Description of package contents (e.g. PPE, wipes, contaminated parts);
(g) Conditioning method(s) used;
(h) Radionuclide(s);
(i) Activity (activities);
(j) Dose rate contact: maximum rate on contact;
(k) Dose rate @ 1 m: rate at 1 m from the container;
(l) Surface contamination;
(m) Mass;
(n) Photographs (container loading and contents, container).

9. STORAGE AND DISPOSAL OF CONDITIONED SOURCES AND WASTE

9.1. STORAGE

The IAEA's Safety Glossary [43] defines storage as "the holding of *radioactive sources*...or *radioactive waste* in a *facility* that provides for their/its *containment*, with the intention of retrieval." Though not part of the definition, Ref. [43] also notes that:

"*Storage* is by definition an interim measure, and the term **[*interim storage*]** would therefore be appropriate only to refer to short term temporary *storage* when contrasting this with the longer term fate of the *waste*."

In the context of Fig. 5, the removal, pre-treatment and conditioning stages would involve interim storage where packages are temporarily waiting for transfer to the next stage of management. Packages containing unconditioned RLCs or waste need to be stored separately from packages containing conditioned RCLs, waste or other conditioned DSRSs or waste in order to avoid possible cross-contamination during an extended storage time.

All of the packaged drums need to be monitored to ensure that they are free from contamination and that the dose rates are consistent with the local requirements for storage.

Storage implies that the waste will be disposed of in the future — storage is not a long term, end point solution for radionuclides with half-lives >10 years. If no disposal facility is yet available, long term storage may be required. For this purpose, the storage has to be properly authorized (licensed) on the basis of a safety case and supporting safety assessments and will need appropriate management and maintenance.

Specific guidance on the storage of small amounts of radioactive waste at different stages of its management is provided in No. SSG-45 [41]. More detailed recommendations are provided in No. WS-G-6.1 [42].

Disused RLCs have been collected and are currently at national storage facilities awaiting further action (conditioning, disposal) in several countries, for example Uruguay [46], North Macedonia [47], Bosnia and Herzegovina [48], Croatia [49], Cuba [50], Brazil [23], China [51], Cyprus [52], France [53], Greece [54], Luxemburg [55], Montenegro [56], Portugal [57], Serbia [58] and Slovenia [59].

As advised in Guidance on the Management of Disused Radioactive Sources [60], the disused sources (recovered from the RLCs) to be stored in a long term storage facility have to be properly conditioned, as required by the regulatory body, and comply with applicable acceptance criteria.

9.2. STORAGE FACILITY

Guidance on the safety of radioactive waste storage facilities is given in No. WS-G-6.1, [42]. Guidance on specific technical details for storage facilities for DSRSs is included in Ref. [4]. The advice given below is mainly for countries where a radioactive waste storage facility is not yet available but where there is a need to establish one at least for their removed RLCs.

The facility will (in accordance with the national legal and regulatory framework) be authorized by the regulatory authority and have signage on the facility such that it is obvious that radioactive materials are contained within. The information also has to detail whom to contact in the event of an emergency. The signage ought to withstand the local weather conditions for the expected duration of the facility and needs to be inspected regularly to ensure that it remains suitable and sufficient.

The facility has to incorporate safety and security measures that provide for the physical protection of the stored sources and waste, such as:

(a) Flooring of sufficient strength to handle the heavy weights associated with cement and lead drums and be easy to decontaminate;
(b) Sufficient height above any probable flooding levels;
(c) Protection against package degradation during storage;
(d) Barriers to prevent easy entry, for example barred windows and secure doors;
(e) Periodic monitoring and maintenance, as appropriate (the frequency of monitoring will depend on the types and activities of radionuclides stored, the robustness of the storage facility and the general security situation);
(f) Procedures for mitigation of problems (such as contamination caused by leaking package).

If the facility is multipurpose, it needs separate areas for the receipt, conditioning and storage of the RLC waste. It needs sufficient room to be able to provide adequate segregation of the sources and any wastes generated from the conditioning process.

If an approved radioactive waste storage facility is not available, possible options to establish an interim storage facility quickly are the use of freight containers (see Fig. 9). ISO freight containers are widely available throughout the world. They have been used as radioactive transport packages and as final disposal containers. They can also be used as storage modules, supplemented with safety and security measures, as required by the regulatory authority (such as those listed above).

ISO freight containers are a flexible, modular low cost method of providing a weatherproof enclosure for waste storage. Being portable, they are flexible in their location and can be relocated if required. They can accommodate a wide range of waste package sizes and weights. The model in Fig. 9 shows a standard 200 l drum inside an ISO freight container properly licensed as storage facility for DSRSs. With the arrangement shown, periodic inspections of the waste packages can be carried out easily.

9.3. DISPOSAL

Radioactive waste from RLCs will be disposed of either in existing facilities or in newly developed ones. Development of a disposal facility requires relevant infrastructure, including a national waste management policy and corresponding strategy for implementation [61], a regulatory framework, and the financial, human and technical resources needed for such a development. If Member States lack resources or the infrastructure to implement RLC disposal, they may seek assistance from other Member States or the IAEA.

The IAEA Safety Glossary [43] defines disposal as the "Emplacement of *waste* in an appropriate *facility* without the intention of retrieval". Options for disposal are near surface disposal or geological disposal. Disposal in boreholes may also be an option [62]. The regulatory and technical aspects of different disposal

a)

b)

FIG. 9. (a) ISO freight containers and (b) an example of waste packages (drum) storage inside them. Credit: IAEA.

options are beyond the scope of this publication. Guidance on the selection of the appropriate disposal option for a given type of radioactive waste, site selection, site investigation, safety and environment assessment, design, licensing, operation, closure and monitoring of disposal facilities are given in a number of IAEA safety standards and technical publications. *waste* in an appropriate *facility* without the intention of retrieval". Options for disposal are near surface disposal or geological disposal. Disposal in boreholes may also be an option [62]. The regulatory and technical aspects of different disposal options are beyond the scope of this publication. Guidance on the selection of the appropriate disposal option for a given type of radioactive waste, site selection, site investigation, safety and environment assessment, design, licensing, operation, closure and monitoring of disposal facilities are given in a number of IAEA safety standards and technical publications.

For existing disposal facilities, waste acceptance criteria (WAC) approved by a regulatory body will determine which waste can be readily accepted for disposal. In the event that RLC waste represents a waste stream not originally planned for at that facility, and its characteristics are not consistent with the WAC, then its acceptability for disposal might be justified by a targeted safety assessment and approved by the regulatory authority.

9.4. RECORD KEEPING

Information collected during storage or disposal of RLCs, their sources and secondary wastes is summarized below (the fifth set of IDB data). All information has to be retained in duplicate at separate locations.

9.4.1. Packages with heads and sources or just sources

(a) Conditioned waste package ID;
(b) Date stored or date disposed;
(c) Storage or disposal location (facility, coordinates).

9.4.2. Packages with secondary waste

(a) Conditioned waste package ID;
(b) Date stored or date disposed;
(c) Storage or disposal location (facility, coordinates).

10. CONCLUSIONS

When RLCs are installed, their Category 4 and 5 (lowest risk) radioactive sources pose little risk to the public.

The risks associated with RLCs are not related to their intended use to promote lightning strikes; rather, over time, the risk to the public increases due to, for example:

(a) Environmental damage caused by wind, rain and lightning strikes;
(b) Little or no maintenance;
(c) Removal without knowledge that they contain SRSs;
(d) Removal by untrained persons, even though they are known to contain SRSs;
(e) Damaged or degraded RLCs, which can lead to radioactive contamination (normally close to the base of the RLC);

(f) Orphaned RLCs, which can end up in scrap metal recycling or conventional landfill sites.

Since the risks of using RLCs outweigh the benefits, their manufacture and installation has been terminated worldwide.

Most countries have programmes or plans to remove RLCs from the public domain, but RLC numbers and locations are not well known, since there was little control when they were first installed.

This publication suggests that:

(a) At a minimum, Member States with RLCs implement public awareness programmes as a first step to compiling a comprehensive inventory of RLCs in their territories to determine the magnitude of the problem;
(b) If Member States are uncertain whether or not they have RLCs, they undertake an assessment to determine the status;
(c) Member States without formal programmes to remove RLCs from the public domain in their territories implement such programmes;
(d) If Member States lack resources or the infrastructure to implement RLC removal programmes, they seek assistance from other Member States or the IAEA;
(e) Member States consult the IAEA Safety Standards and Security Series publications and Technical Documents on the management of disused SRSs (DSRSs), since they are related to the management of sources from RLCs;
(f) If disposal of DRSRs within lead containing RLC heads is planned, consideration needs to be given to the waste disposal facility's acceptance criteria for lead containing waste.

Appendix I

EXAMPLES OF RADIOACTIVE LIGHTNING CONDUCTORS

The IAEA has developed and implemented a publicly available database that includes a variety of radioactive lightning conductor models (Fig. 10) [63]. One of the purposes of the RLC database is to present several models of RLC and provide details on each model's construction and the types of sealed sources they contain[5].

Examples of different models of radioactive lightning conductors containing alpha, beta or gamma emitting radionuclides are show in Figs 11–38.

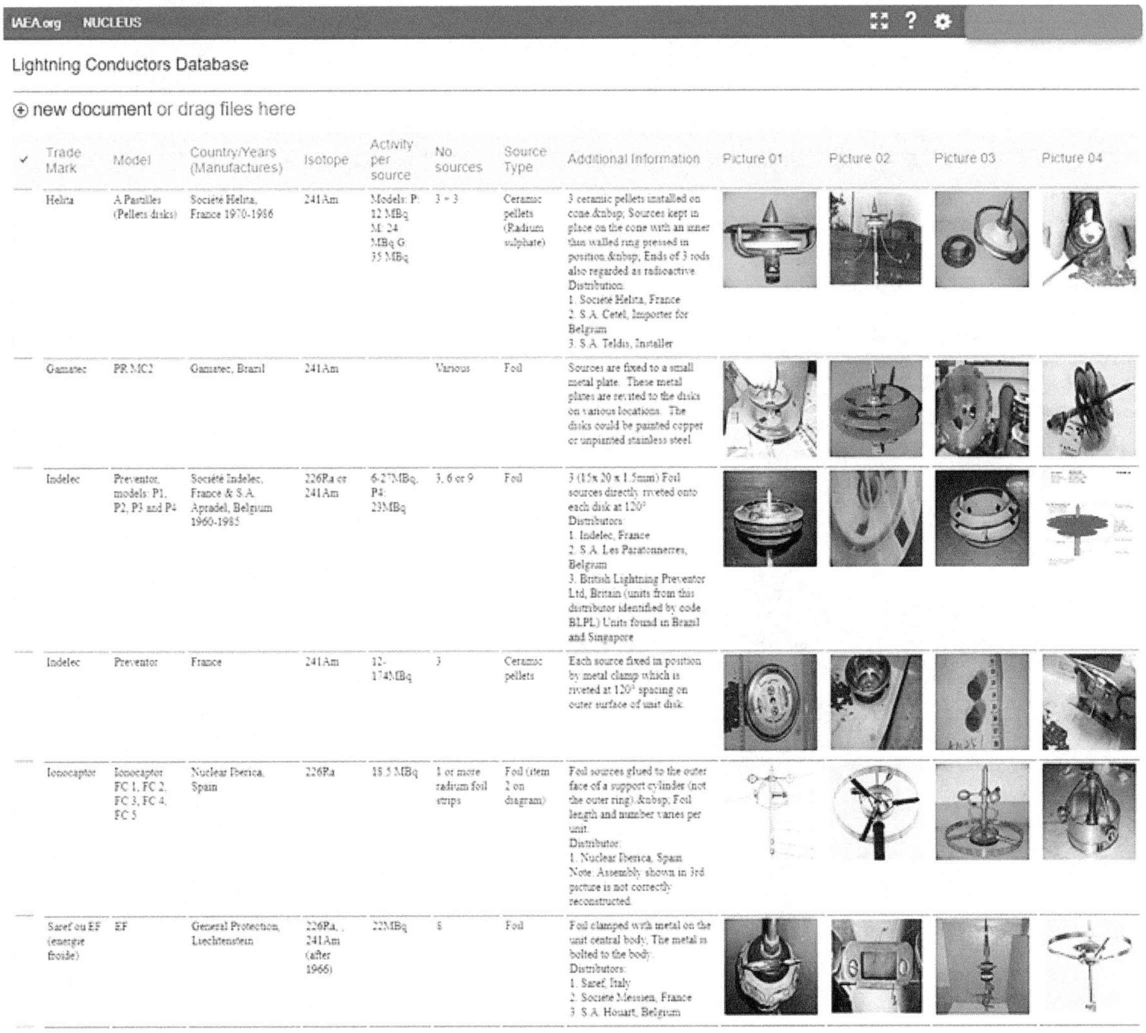

IAEA.org NUCLEUS

Lightning Conductors Database

⊕ new document or drag files here

✓	Trade Mark	Model	Country/Years (Manufactures)	Isotope	Activity per source	No sources	Source Type	Additional Information	Picture 01	Picture 02	Picture 03	Picture 04
	Helita	A Pastilles (Pellets disks)	Société Helita, France 1970-1986	^{241}Am	Models: P: 12 MBq M: 24 MBq G: 35 MBq	3 + 3	Ceramic pellets (Radium sulphate)	3 ceramic pellets installed on cone Sources kept in place on the cone with an inner thin walled ring pressed in position Ends of 3 rods also regarded as radioactive. Distribution: 1. Société Helita, France 2. S.A. Cetel, Importer for Belgium 3. S.A. Teldis, Installer				
	Gamatec	PR MC2	Gamatec, Brazil	^{241}Am		Various	Foil	Sources are fixed to a small metal plate. These metal plates are revited to the disks on various locations. The disks could be painted copper or unpainted stainless steel				
	Indelec	Preventor, models: P1, P2, P3 and P4	Société Indelec, France & S.A. Apradel, Belgium 1960-1985	^{226}Ra or ^{241}Am	6-27MBq, P4: 23MBq	3, 6 or 9	Foil	3 (15x 20 x 1.5mm) Foil sources directly riveted onto each disk at 120° Distributors: 1. Indelec, France 2. S.A. Les Paratonnerres, Belgium 3. British Lightning Preventor Ltd, Britain (units from this distributor identified by code BLPL) Units found in Brazil and Singapore				
	Indelec	Preventor	France	^{241}Am	12-174MBq	3	Ceramic pellets	Each source fixed in position by metal clamp which is riveted at 120° spacing on outer surface of unit disk.				
	Ionocaptor	Ionocaptor FC 1, FC 2, FC 3, FC 4, FC 5	Nuclear Iberica, Spain	^{226}Ra	18.5 MBq	1 or more radium foil strips	Foil (item 2 on diagram)	Foil sources glued to the outer face of a support cylinder (not the outer ring) Foil length and number varies per unit. Distributor: 1. Nuclear Iberica, Spain Note: Assembly shown in 3rd picture is not correctly reconstructed.				
	Saref ou EF (energie froide)	EF	General Protection, Liechtenstein	^{226}Ra, , ^{241}Am (after 1966)	22MBq	8	Foil	Foil clamped with metal on the unit central body. The metal is bolted to the body. Distributors: 1. Saref, Italy 2. Societe Messien, France 3. S.A. Houart, Belgium				

FIG. 10. Screen capture of the RLC database within the IAEA Professional Network DSRSNet.

[5] See: https://nucleus.iaea.org/sites/connect-members/dsrs/Pages/Lightening-Conductors.aspx

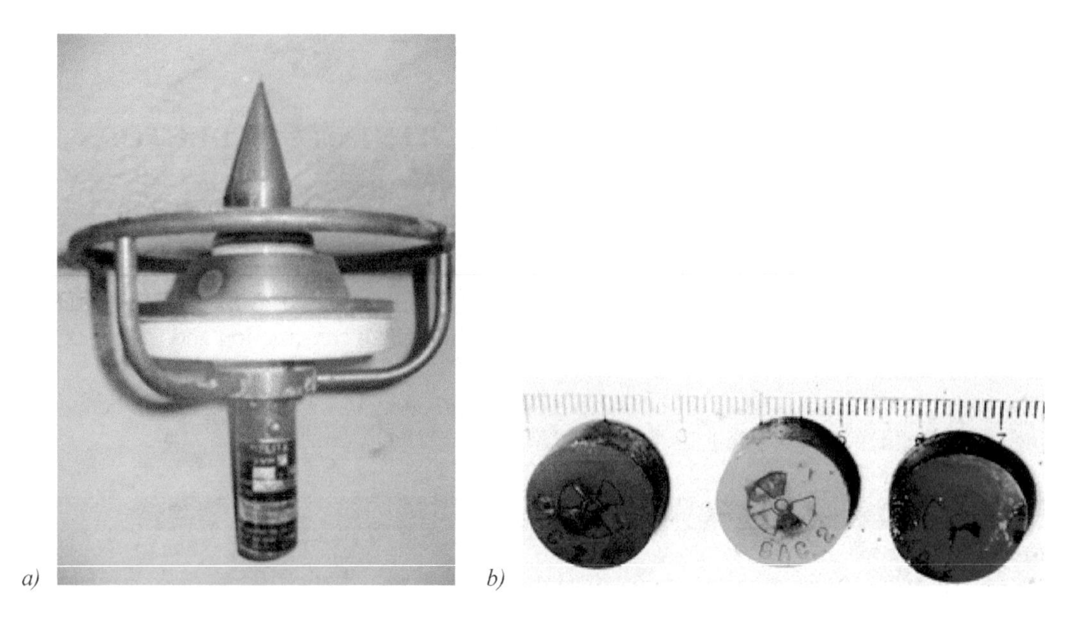

FIG. 11. (a) RLC, Helita, model AMH (pellets/disks). (b) ^{241}Am sources (model SAC 2).

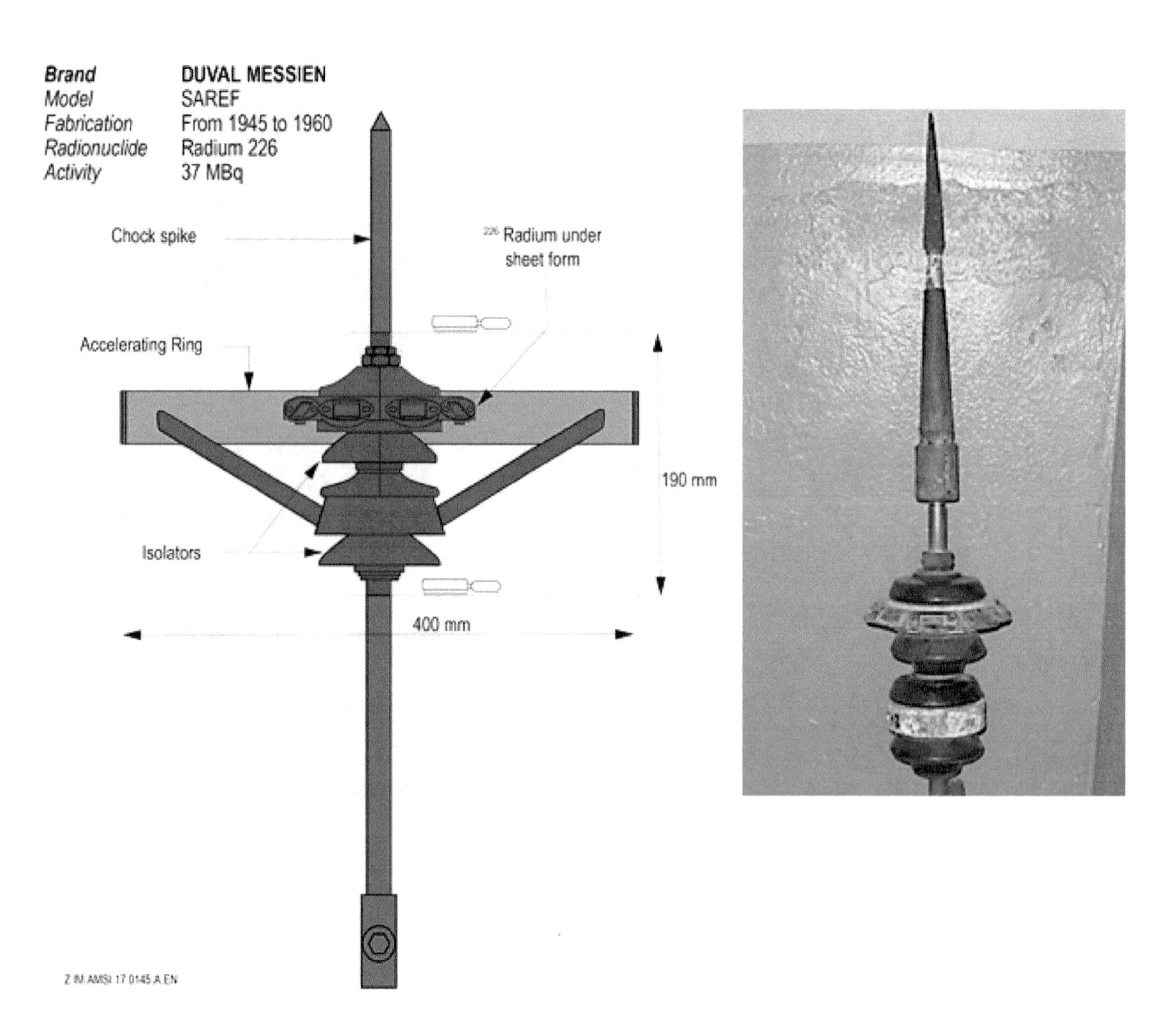

FIG. 12. RLC, Duval Messien, model SAREF, ^{226}Ra and ^{241}Am sources. Drawing courtesy of ANDRA.

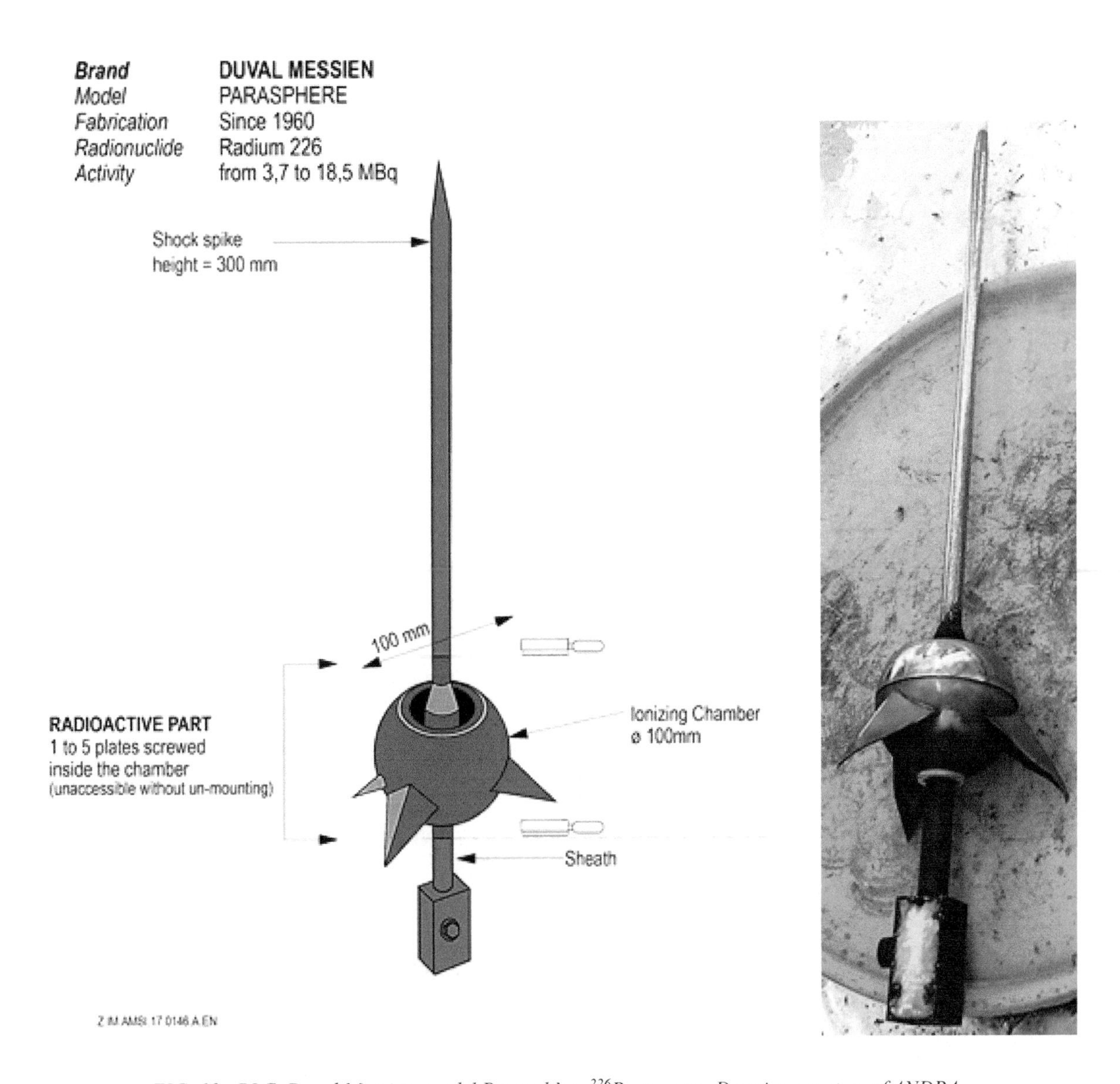

FIG. 13. RLC, Duval Messien, model Parasphère, ^{226}Ra sources. Drawing courtesy of ANDRA.

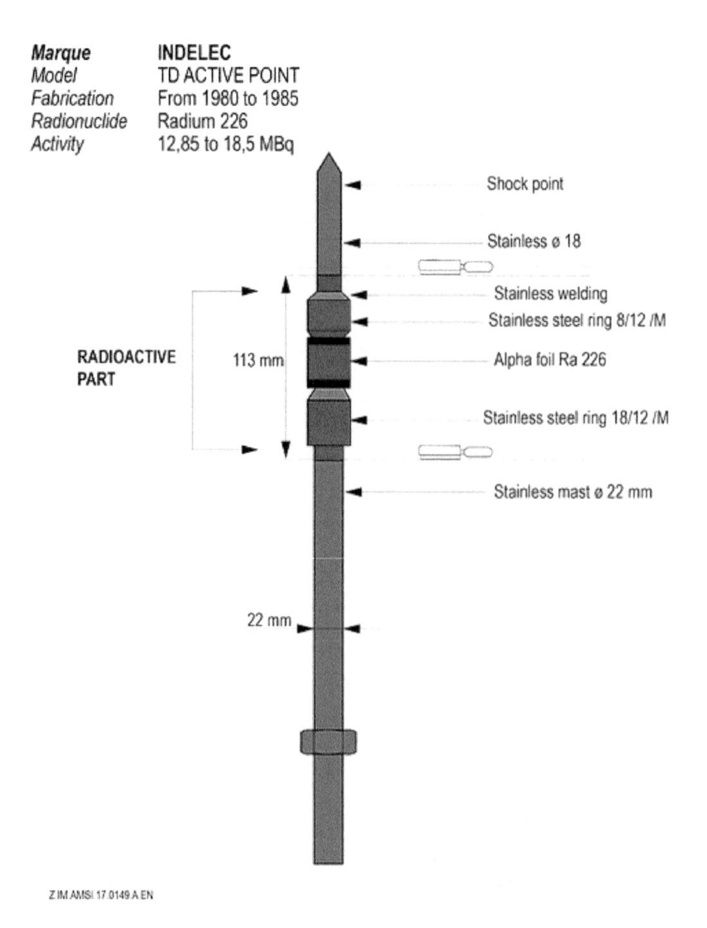

FIG. 14. RLC, Indelec, model TD Active Point, ^{226}Ra sources. Courtesy of ANDRA.

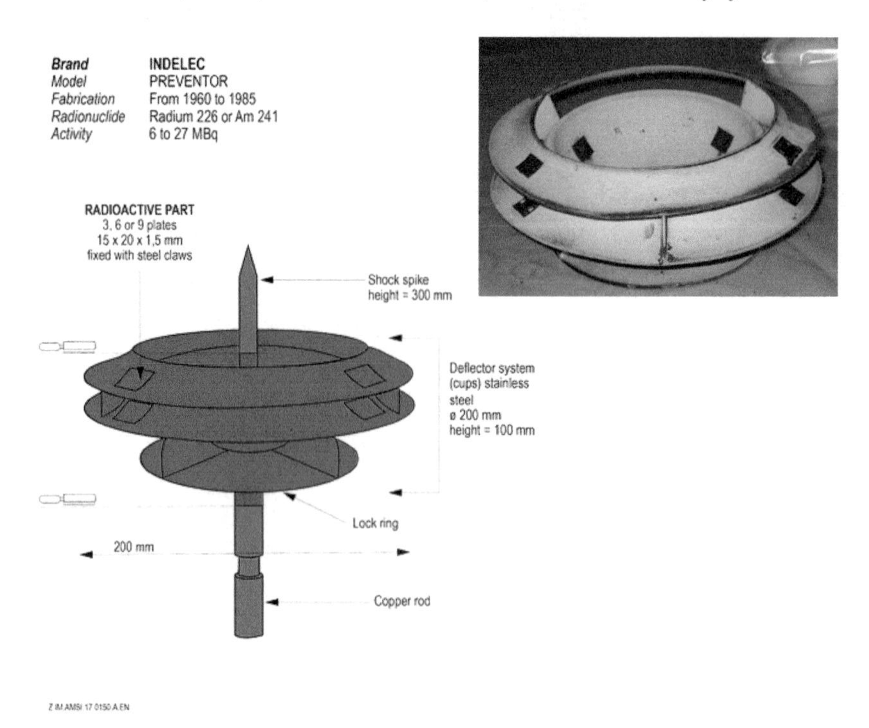

FIG. 15. RLC, Indelec, model Preventor (P1, P2, P3 and P4), ^{241}Am and ^{226}Ra sources. Drawing courtesy of ANDRA.

FIG. 16. RLC, unidentified model (similar to Preventor), six ^{226}Ra sources.

FIG. 17. RLC, Indelec (distributor), model Preventor, ^{241}Am sources. Courtesy of W.T. Hoo, National Environment Agency, Singapore.

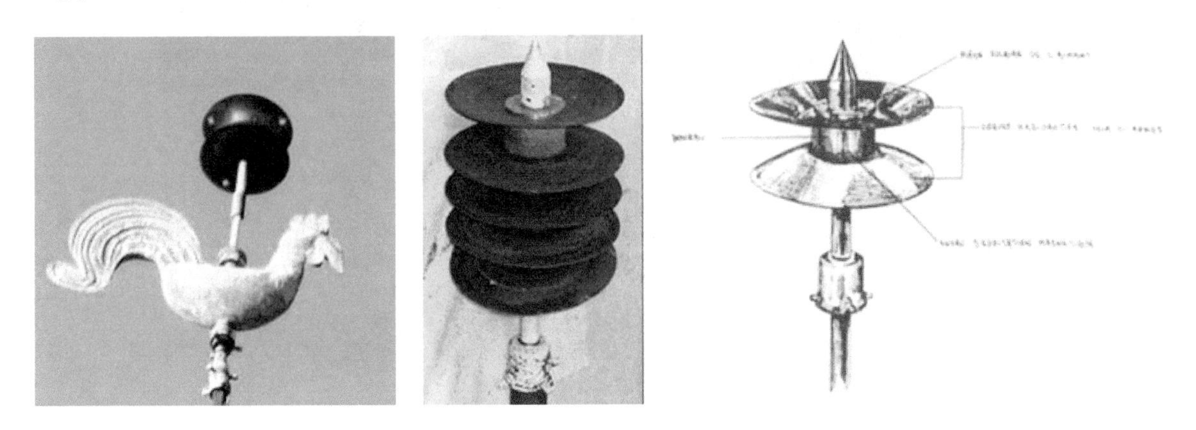

FIG. 18. RLC, Kapton, model 1949, ^{226}Ra sources, made in Belgium. Courtesy of FANC, Belgium.

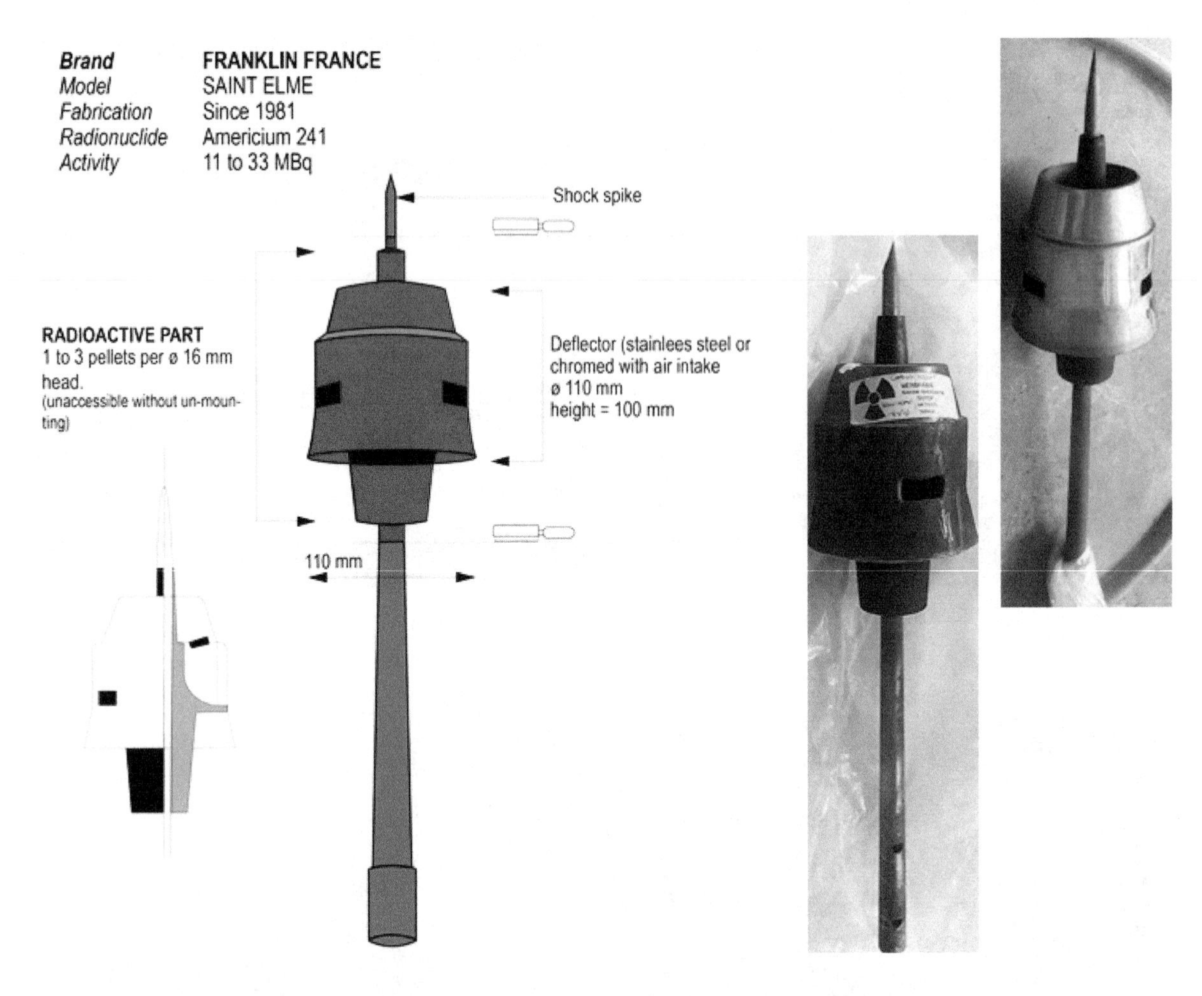

FIG. 19. RLC, Franklin France, model Saint Elme, ^{241}Am sources. Drawing courtesy of ANDRA.

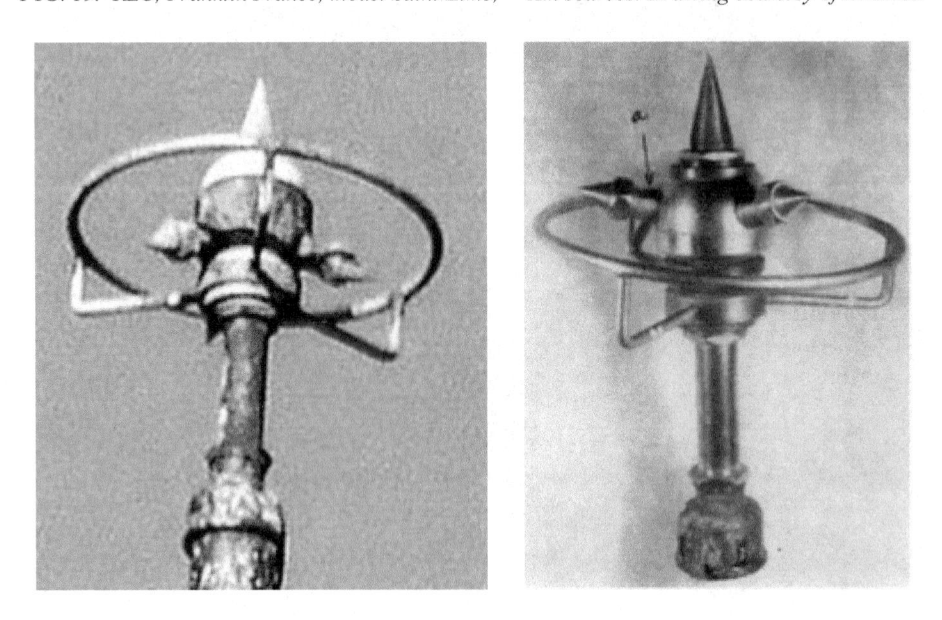

FIG. 20. RLC, Combinator, models SC A, SC 1, SC 2, SC 3, SC 4, SC 5, SC 6 and SC 350, ^{226}Ra sources, made in Belgium.

FIG. 21. RLC, Ionox, ^{226}Ra or ^{241}Am sources.

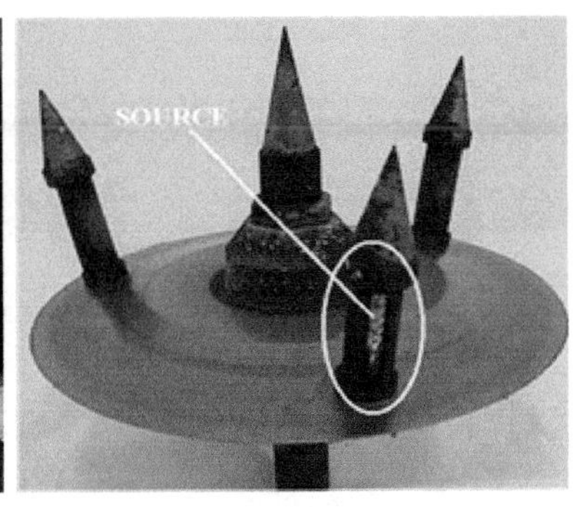

FIG. 22. RLC, unidentified model (similar to Horemans-Souply, Ionix and Combinator), ^{226}Ra or ^{241}Am sources.

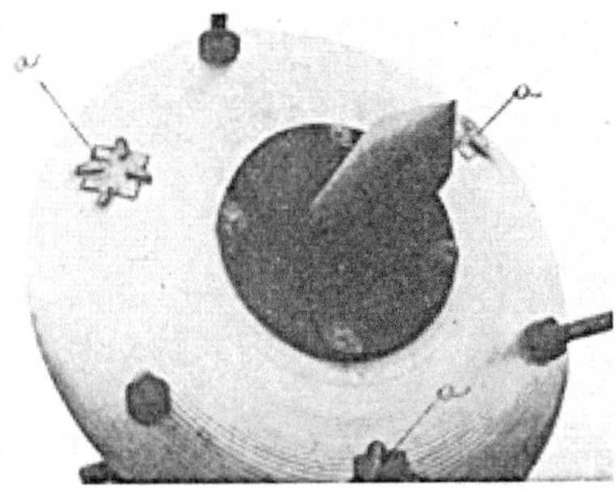

FIG. 23. RLC, Horemans-Souply, ^{226}Ra sources, made in Belgium. Courtesy FANC.

FIG. 24. RLC, Fair Raythor, model Minor, ^{226}Ra or ^{241}Am sources.

FIG. 25. RLC, Fair Raythor, model Mayor, ^{226}Ra or ^{241}Am sources.

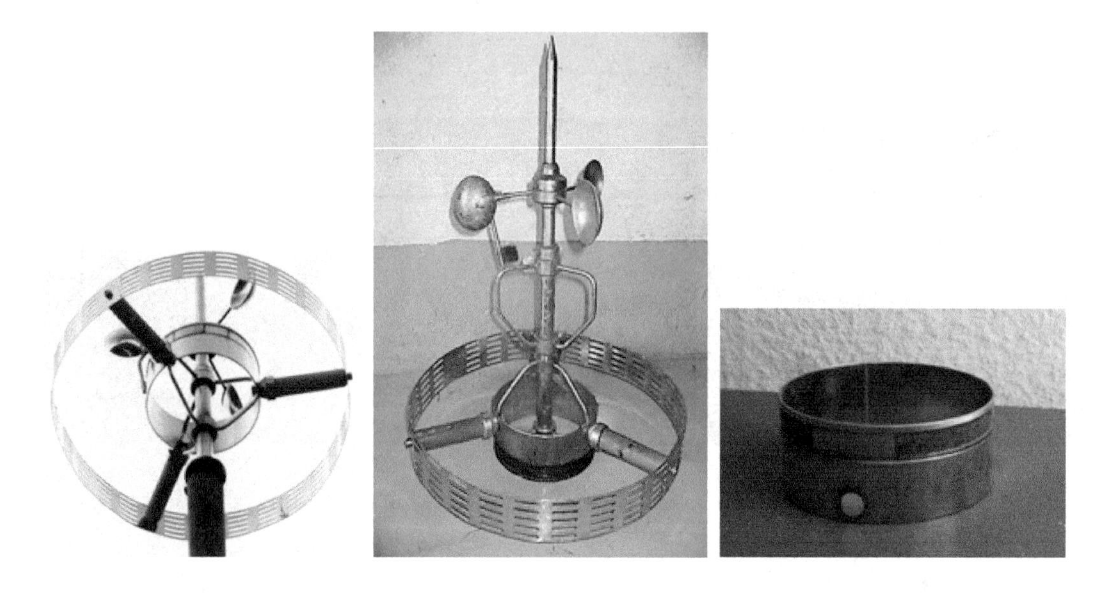

FIG. 26. RLC, Nuclear Ibérica, models Ionocaptor FC 1, FC 2, FC 3, FC 4 and FC 5, ^{241}Am or ^{226}Ra sources, made in Spain.

FIG. 27. RLC, Amerion (Brazil), models R-25, R-50, R-75 and R-100, ^{241}Am sources.

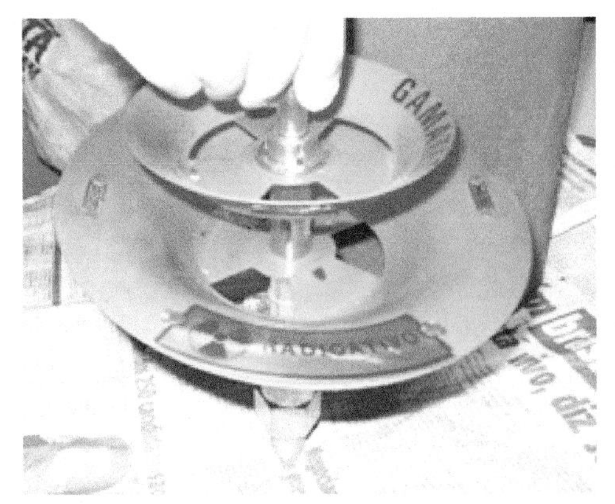

FIG. 28. RLC, Gamatec (Brazil), model PR MC2, ^{241}Am sources.

FIG. 29. RLC, model unknown, made in Brazil, ^{241}Am sources.

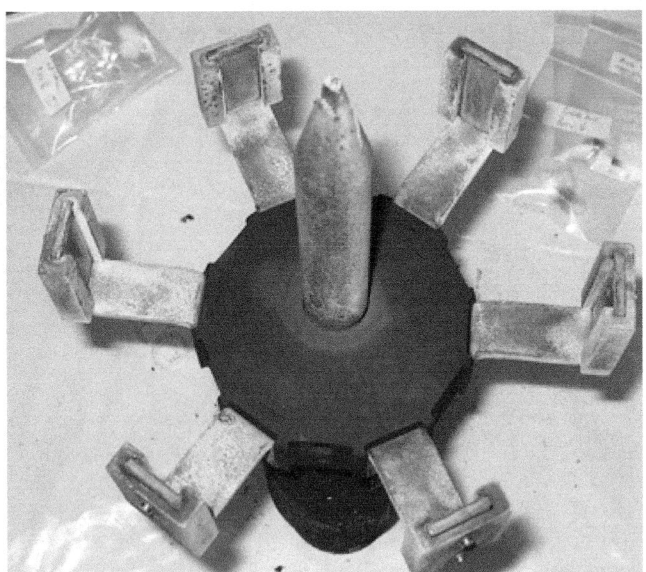

FIG. 30. RLC, E.F. Australasia Pty Ltd, model EF33, ^{241}Am sources.

FIG. 31. RLC, Nuclear Ibérica, model Minocaptor, ^{241}Am sources.

FIG. 32. RLC, Energía fría, ^{14}C, ^{226}Ra or ^{241}Am sources.

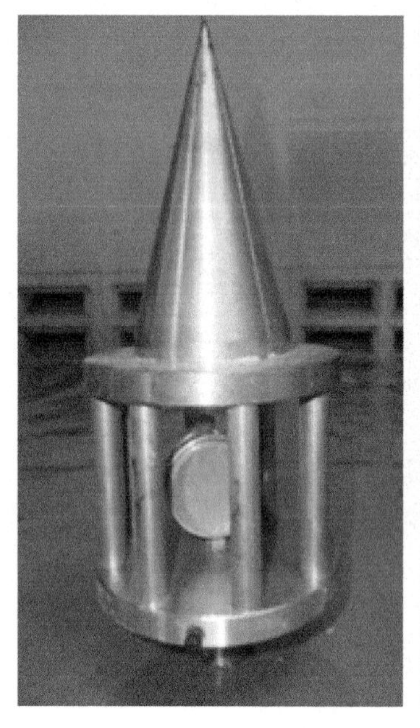

FIG. 33. RLC, Radac, ^{85}Kr sources.

FIG. 34. RLC, Radiber, ^{90}Sr sources.

FIG. 35. RLC, ELIND Valjevo Serbia, model unknown, ^{152}Eu and ^{60}Co sources.

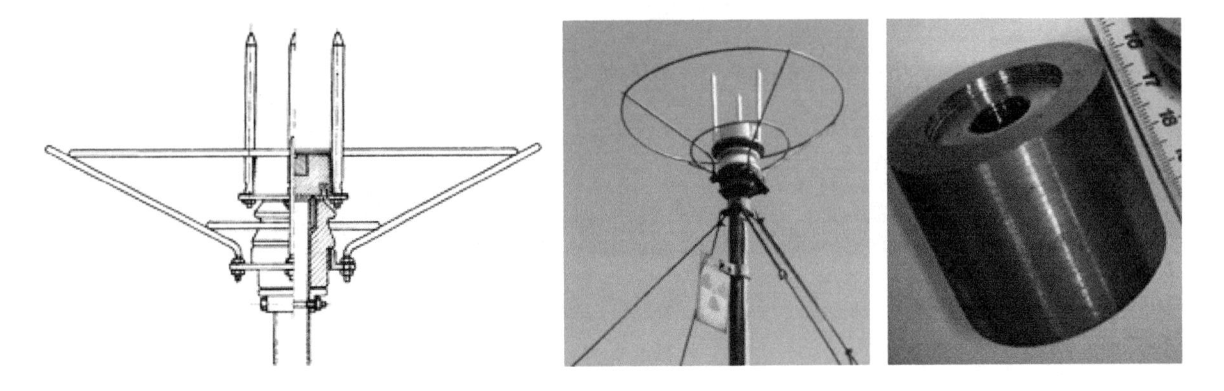

FIG. 36. RLC, ELIND Valjevo Serbia, model unknown, ^{152}Eu and ^{154}Eu sources.

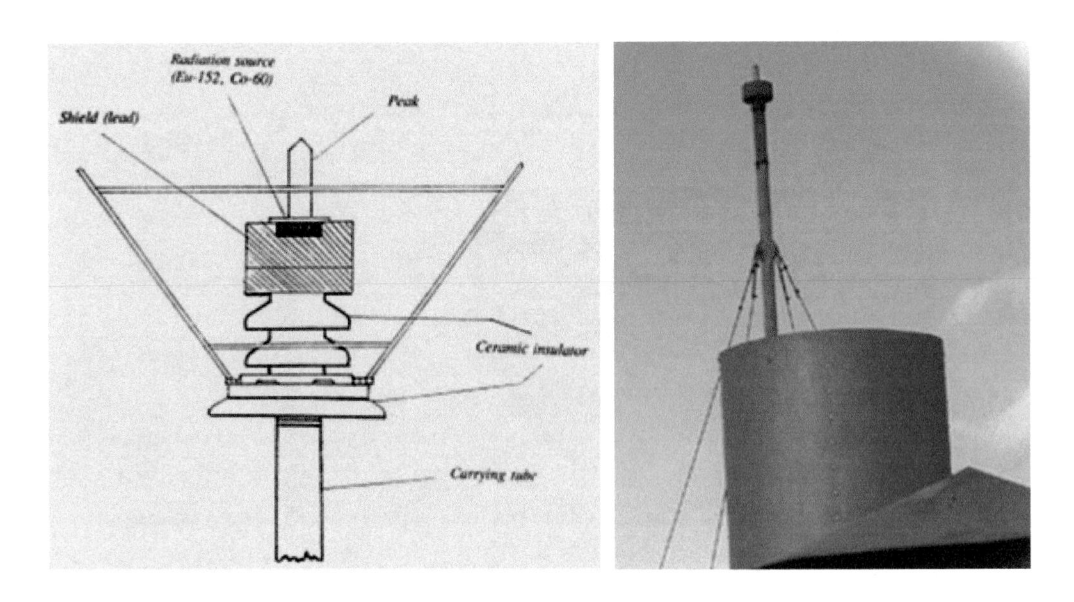

FIG. 37. RLC, GOI Zagreb, Croatia, model unknown, ^{152}Eu and ^{154}Eu sources.

FIG. 38. RLC, Goran Produkt Vrbovsko Croatia and Slavija Elektro Belgrade, model unknown, ^{152}Eu and ^{154}Eu sources.

Appendix II

EXAMPLE OF A PUBLIC AWARENESS PROGRAMME

An example of a public awareness programme can be seen in Fig. 39.

[1] Attention: RLCs in the region. They have been banned since 1987 but they are still on our roofs.
[2] The potential risks to your well being. [3] How RLCs are removed. [4] How RLCs are recognized.

FIG. 39. Example of an RLC awareness fact sheet in Belgium. Courtesy of FANC, Belgium.

REFERENCES

[1] EUROPEAN COMMISSION, FOOD AND AGRICULTURE ORGANIZATION OF THE UNITED NATIONS, INTERNATIONAL ATOMIC ENERGY AGENCY, INTERNATIONAL LABOUR ORGANIZATION, OECD NUCLEAR ENERGY AGENCY, PAN AMERICAN HEALTH ORGANIZATION, UNITED NATIONS ENVIRONMENT PROGRAMME, WORLD HEALTH ORGANIZATION, Radiation Protection and Safety of Radiation Sources: International Basic Safety Standards, IAEA Safety Standards Series No. GSR Part 3, IAEA, Vienna (2014).

[2] INTERNATIONAL ATOMIC ENERGY AGENCY, Nature and Magnitude of the Problem of Spent Radiation Sources, IAEA-TECDOC-620, IAEA, Vienna (1991).

[3] INTERNATIONAL ATOMIC ENERGY AGENCY, Methods to Identify and Locate Spent Radiation Sources, IAEA-TECDOC-804, IAEA, Vienna (1995).

[4] INTERNATIONAL ATOMIC ENERGY AGENCY, Management of Disused Sealed Radioactive Sources, Nuclear Energy Series No. NW-T-1.3, IAEA, Vienna (2014).

[5] INTERNATIONAL ATOMIC ENERGY AGENCY, Handling, Conditioning and Storage of Spent Sealed Radioactive Sources, IAEA-TECDOC-1145, IAEA, Vienna (2000).

[6] INTERNATIONAL ATOMIC ENERGY AGENCY, Management for the Prevention of Accidents from Disused Sealed Radioactive Sources, IAEA-TECDOC-1205, IAEA, Vienna, (2001).

[7] INTERNATIONAL ATOMIC ENERGY AGENCY, Management of Disused Long Lived Sealed Radioactive Sources (LLSRS), IAEA-TECDOC-1357, IAEA, Vienna (2003).

[8] FRENCH NATIONAL RADIOACTIVE WASTE MANAGEMENT AGENCY (ANDRA), La Récupération des Paratonnerres Radioactifs en France, Fiche 3, ANDRA, Paris (1995).

[9] HARTONO, Z.A, ROBIAH, I., Conventional and Un-conventional Lightning Air Terminals: An Overview, Forum on Lightning Protection, Kuala Lumpur (2004).

[10] UMAN, M.A., RAKOV, V.A., A critical review of nonconventional approaches to lightning protection, Bull. Amer. Meteor. Soc. 83 (2002) 1809–1820.

[11] DARVENIZA, M., MACKERRAS, D., LIEW, A.C., Standard and non-standard lightning protection methods, Aust. J. Electr. 7 (1987) 133–140.

[12] EUROPEAN COMMISSION, A Review of Consumer Products Containing Radioactive Substances in the European Union, Radiation Protection 146, Final Report of the Study Contract for the European Commission, B4-3040/2001/327150/MAR/C4, EC, Brussels (2007).

[13] INTERNATIONAL ATOMIC ENERGY AGENCY, Categorization of Radioactive Sources, IAEA Safety Standards Series No. RS-G-1.9, IAEA, Vienna (2005).

[14] INTERNATIONAL ATOMIC ENERGY AGENCY, Review of Sealed Source Designs and Manufacturing Techniques Affecting Disused Source Management, IAEA-TECDOC-1690, IAEA, Vienna (2012).

[15] INTERNATIONAL ATOMIC ENERGY AGENCY, Conditioning and Interim Storage of Spent Radium Sources, IAEA-TECDOC-886, IAEA, Vienna (1996).

[16] COMMITTEE ON RADIOACTIVE SOURCE USE AND REPLACEMENT, NATIONAL RESEARCH COUNCIL NATIONAL ACADEMY OF SCIENCES, Radiation Source Use and Replacement: Abbreviated Version, National Academy of Sciences, Washington, DC (2008).

[17] INTERNATIONAL ATOMIC ENERGY AGENCY, Control of Orphan Sources and other Radioactive Material in the Metal Recycling and Production Industries, IAEA Safety Standards Series No. SSG-17, IAEA, Vienna (2012).

[18] Novakovic, M., Incident involving radioactive lightning conductors in Croatia, European ALARA Newsletter 18 9 (2006).

[19] INTERNATIONAL ATOMIC ENERGY AGENCY, Hazardous Manipulation with a Lightning Rod with Ra-226 (2011),
https://www-news.iaea.org/ErfView.aspx?mId=f238a1e1-7364-45b2-af78-00a27fecea71

[20] CIRAJ-BJELAC, O., et. al., A radiological incident with a radioactive lightning rod source found in a vehicle used by film crew members, Radiat. Prot. Dosimetry 141 (2010) 309–314.

[21] INTERNATIONAL ATOMIC ENERGY AGENCY, Justification of Practices, Including Non-Medical Human Imaging, IAEA Safety Standards Series No. GSG-5, IAEA, Vienna (2014).

[22] RAELE, M.P., et al., "Laser decontamination of the radioactive lightning conductors", ISRP 2012 (Proc. Int. Symp. on Radiation Physics, Rio de Janeiro, 2012), International Radiation Physics Society, Rio de Janeiro (2012).

[23] FEDERATIVE REPUBLIC OF BRAZIL, National Report of Brazil for the 6th Review Meeting, Joint Convention on the Safety of Spent Fuel Management and on the Safety of Radioactive Waste Management, Brazil (2017).

[24] FRENCH NATIONAL RADIOACTIVE WASTE MANAGEMENT AGENCY (ANDRA), Managing Diffuse Nuclear Waste, Essential Series 209 VA, ANDRA, Paris (2007).

[25] DEPARTMENT OF RADIATION PROTECTION, National Report on the Measures Taken by Luxembourg to Fulfill the Obligations Laid Down in the Joint Convention on the Safety of Spent Fuel Management and on the Safety of Radioactive Waste Management, Fourth Review Meeting of the Contracting Parties in 2012, DRP, Luxembourg (2012).

[26] EMPRESA NACIONAL DE RESIDUOS RADIACTIVOS, Activities and Projects/Radioactive Waste Management Campaigns/Radioactive Lightning Conductors (2009),
http://www.enresa.es/activities_and_projects/other_activities/radioactive_lightning_conductors

[27] AMPER ELEKTRIK, The Radioactive Lightning Conductors (2016),
http://www.amper.com.tr/en/products/external-lightning-systems

[28] EUROPEAN ATOMIC ENERGY COMMUNITY, FOOD AND AGRICULTURE ORGANIZATION OF THE UNITED NATIONS, INTERNATIONAL ATOMIC ENERGY AGENCY, INTERNATIONAL LABOUR ORGANIZATION, INTERNATIONAL MARITIME ORGANIZATION, OECD NUCLEAR ENERGY AGENCY, PAN AMERICAN HEALTH ORGANIZATION, UNITED NATIONS ENVIRONMENT PROGRAMME, WORLD HEALTH ORGANIZATION, Fundamental Safety Principles, IAEA Safety Standards Series No. SF-1, IAEA, Vienna (2006).

[29] INTERNATIONAL ATOMIC ENERGY AGENCY, Governmental, Legal and Regulatory Framework for Safety, IAEA Safety Standards Series No. GSR Part 1 (Rev. 1), IAEA, Vienna (2016).

[30] INTERNATIONAL ATOMIC ENERGY AGENCY, Predisposal Management of Radioactive Waste, IAEA Safety Standards Series No. GSR Part 5, IAEA, Vienna (2009).

[31] INTERNATIONAL ATOMIC ENERGY AGENCY, Security of Radioactive Sources, IAEA Nuclear Security Series No. 11, IAEA, Vienna (2009).

[32] EUROPEAN COMMISSION, Council Directive 11/70 Euratom of 19 July 2011, Establishing a Community Framework for the Responsible and Safe Management of Spent Fuel and Radioactive Waste, Official Journal of the European Communities No. L 199/48, Office for Official Publications of the European Communities, Luxembourg (2011).

[33] INTERNATIONAL ATOMIC ENERGY AGENCY, Leadership and Management for Safety, IAEA Safety Standards Series No. GSR Part 2, IAEA, Vienna (2016).

[34] INTERNATIONAL ATOMIC ENERGY AGENCY, The Management System for Facilities and Activities, IAEA Safety Standards Series No. GS-R-3, IAEA, Vienna (2006).

[35] INTERNATIONAL ATOMIC ENERGY AGENCY, Application of the Management System for Facilities and Activities, IAEA Safety Standards Series No.GS-G-3.1, IAEA, Vienna (2006).

[36] INTERNATIONAL ATOMIC ENERGY AGENCY, Leadership, Management and Culture for Safety in Radioactive Waste Management, IAEA Safety Standards Series No. GSG-16, IAEA, Vienna (2022).

[37] INTERNATIONAL ATOMIC ENERGY AGENCY, National Strategy for Regaining Control over Orphan Sources and Improving Control over Vulnerable Sources, IAEA Safety Standards Series No. SSG-19, IAEA, Vienna (2009).

[38] INTERNATIONAL ATOMIC ENERGY AGENCY, Workplace Monitoring for Radiation and Contamination, Practical Radiation Technical Manual No. 1 (Rev. 1), IAEA, Vienna (2004).

[39] INTERNATIONAL ATOMIC ENERGY AGENCY, Regulations for the Safe Transport of Radioactive Material, IAEA Safety Standards Series No. SSR-6 (Rev. 1), IAEA, Vienna (2018).

[40] INTERNATIONAL ATOMIC ENERGY AGENCY, Predisposal Management of Radioactive Waste from Nuclear Power Plants and Research Reactors, IAEA Safety Standards Series No. SSG-40, IAEA, Vienna (2016).

[41] INTERNATIONAL ATOMIC ENERGY AGENCY, Predisposal Management of Radioactive Waste from the Use of Radioactive Material in Medicine, Industry, Agriculture, Research and Education, IAEA Safety Standards Series No. SSG-45, IAEA, Vienna (2019).

[42] INTERNATIONAL ATOMIC ENERGY AGENCY, Storage of Radioactive Waste, IAEA Safety Standards Series No. WS-G-6.1, IAEA, Vienna (2006).

[43] INTERNATIONAL ATOMIC ENERGY AGENCY, IAEA Safety Glossary: 2018 Edition, Non-serial Publications, IAEA, Vienna (2019).

[44] INTERNATIONAL ATOMIC ENERGY AGENCY, Code of Conduct on the Safety and Security of Radioactive Sources, Non-serial Publications, IAEA, Vienna (2004).

[45] INTERNATIONAL ATOMIC ENERGY AGENCY, Safety of Radiation Generators and Sealed Radioactive Sources, IAEA Safety Standards Series No. RS-G-1.10, IAEA, Vienna (2006).
[46] NATIONAL REGULATORY AUTHORITY IN RADIATION PROTECTION, URUGUAY, National Report for the Sixth Review Meeting, Joint Convention on the Safety of Spent Fuel Management and on the Safety of Radioactive Waste Management, Ministerio de Industria, Energía y Minería, Montevideo (2017)
[47] RADIATION SAFETY DIRECTORATE REPUBLIC OF MACEDONIA, Third National Report, Joint Convention on the Safety of Spent Fuel Management and on the Safety of Radioactive Waste Management, Radiation Safety Directorate, Skopje (2017).
[48] STATE REGULATORY AGENCY FOR RADIATION AND NUCLEAR SAFETY, BOSNIA AND HERZEGOVINA, National Report of Bosnia and Herzegovina on the implementation of the obligations under the Joint Convention on the Safety of Spent Fuel and on the Safety of Radioactive Waste Management, State Regulatory Agency for Radiation and Nuclear Safety, Sarajevo (2017).
[49] STATE OFFICE FOR RADIOLOGICAL AND NUCLEAR SAFETY, 6th National Report on Implementation of the obligations under the Joint Convention on the Safety of Spent Fuel Management and on the Safety of Radioactive Waste Management, State Office for Radiological and Nuclear Safety of Croatia, Zagreb (2017).
[50] NATIONAL NUCLEAR SAFETY CENTER, CENTER FOR RADIATION PROTECTION AND HYGIENE, First National Report of Joint Convention on the Safety of Spent Fuel Management and on the Safety of Radioactive Waste Management, National Nuclear Safety Center, Havana (2017).
[51] THE PEOPLE'S REPUBLIC OF CHINA, Fourth National Report for the Joint Convention on the Safety of Spent Fuel Management and on the Safety of Radioactive Waste Management, Beijing (2017).
[52] RADIATION INSPECTION AND CONTROL SERVICE, CYPRUS, National Report on the Implementation of the Obligations under the Joint Convention on the Safety of Spent Fuel Management and on the Safety of Radioactive Waste Management. Submitted for the purposes of the 6th Review Meeting of the Convention, Ministry of Labour, Welfare and Social Insurance, Department of Labour Inspection, Radiation Inspection and Control Service, Nicosia (2017).
[53] FRENCH NUCLEAR SAFETY AUTHORITY, Sixth National Report on Compliance with the Joint Convention Obligations. Joint Convention on the Safety of Spent Fuel Management and on the Safety of Radioactive Waste Management, ASN, Paris (2017).
[54] GREEK ATOMIC ENERGY COMMISSION, National Report of Greece, Joint Convention on the Safety of Spent Fuel Management and on the Safety of Radioactive Waste Management, Greek Atomic Energy Commission, Athens (2017).
[55] DEPARTMENT OF RADIATION PROTECTION (DRP), National Report on the Measures Taken by Luxembourg to Fulfil the Obligations Laid Down in the Joint Convention on the Safety of Spent Fuel Management and on the Safety of Radioactive Waste Management, Sixth Review Meeting of the Contracting Parties, DRP, Luxembourg (2017).
[56] MINISTRY OF SUSTAINABLE DEVELOPMENT AND TOURISM, MONTENEGRO, Third National Report on the Implementation of the Obligations under the Joint Convention on the Safety of Spent Fuel Management and on the Safety of Radioactive Waste Management, Ministry of Sustainable Development and Tourism, Podgorica (2017).
[57] REGULATORY COMMISSION FOR THE SAFETY OF NUCLEAR INSTALLATIONS, PORTUGAL, Third National Report (2014–2017), 6th Review Meeting of the Contracting Parties, Joint Convention on the Safety of Spent Fuel Management and on the Safety of Radioactive Waste Management, Comissão Reguladora para a Segurança das Instalações Nucleares, Lisbon (2017).
[58] SERBIAN RADIATION PROTECTION AND NUCLEAR SAFETY AGENCY, First National Report, Joint Convention on the Safety of Spent Fuel Management and on the Safety of Radioactive Waste Management, Serbian Radiation Protection and Nuclear Safety Agency, Beograd (2018).
[59] SLOVENIAN NUCLEAR SAFETY ADMINISTRATION, Sixth Slovenian Report under the Joint Convention on the Safety of Spent Fuel Management and on the Safety of Radioactive Waste Management, Slovenian Nuclear Safety Administration, Ljubljana (2017).
[60] INTERNATIONAL ATOMIC ENERGY AGENCY Guidance on the Management of Disused Radioactive Sources, Non-serial Publications, IAEA, Vienna (2018).
[61] INTERNATIONAL ATOMIC ENERGY AGENCY, Policies and Strategies for Radioactive Waste Management, IAEA Nuclear Energy Series No. NW-G-1.1, IAEA, Vienna (2009).
[62] INTERNATIONAL ATOMIC ENERGY AGENCY, Borehole Disposal Facilities for Radioactive Waste, IAEA Safety Standards Series No. SSG-1, IAEA, Vienna (2009).

[63] INTERNATIONAL ATOMIC ENERGY AGENCY, Disused Sealed Radioactive Sources Network — DSRSNet, Radioactive Lightning Conductors Database (2021),
https://nucleus.iaea.org/sites/connect-members/dsrs/Pages/Lightening-Conductors.aspx

Annex I

EXAMPLE OF THE FULL LIFE CYCLE MANAGEMENT OF RLCs REMOVED FROM THE PUBLIC DOMAIN: SPAIN

Several models of RLC exist in Spain, but all of them are minor variants of a circular geometry with a central ring where the source is placed. They have an additional external ring that functions to increase the ionization effect. The tip of the conductor is ~20–25 cm from the central ring. The diameter of the head is ~40 cm, with a maximum size of 50 cm.

Approximately 90% of RLCs in Spain used ^{241}Am, typically in the form of oxide layers that were insoluble and non-volatile and sintered in a noble metal foil.

The various models of RLC include Minocaptor, Energia Fria and Ionocaptor, which contain one of the following radionuclides: ^{90}Sr, ^{85}Kr, ^{14}C and ^{241}Am. In some RLC models (e.g. Preventor, Helita and Energia Fria) the main radionuclide is ^{226}Ra.

The decision to remove RLCs from the public domain was based on two principles of radiation protection: the principle of justification — that is, no major benefit is obtained with the use of RLCs — and the principle of optimization (or ALARA criteria) in which the risk of exposure and contamination is minimized for humans and the environment.

The management alternatives investigated for RLC removal from the public domain were: (a) long term management in Spain and (b) recycling of sources abroad. For ^{241}Am, the recycling alternative was pursued and is described here.

The processes involved with this option were retrieval, conditioning, secondary waste management and recycling of the ^{241}Am abroad.

Retrieval starts with notification of the legal owner and/or the organization in possession of the RLC. Notification describes the role of the owner and/or holder in the retrieval process and the work plan for the removal.

The first operational step is dismounting RLCs from the structures where they are installed. Next, the head with the radioactive source is separated from the rest of the device. Radiological measurements of the area are made and, if needed, the area is decontaminated. Finally, a radiological certificate is issued, certifying that the area is free or radiological contamination.

The separated head is pretreated, for example packaged, to ensure that it can be transported safely to the conditioning facility. At this facility:

(i) All documentation is controlled and quality assured;
(ii) The heads are dismantled and the sources separated from remaining components;
(iii) Secondary waste from dismantling is volume reduced, packed in 200 L drums and characterized non-destructively;
(iv) Foil strips containing multiple ^{241}Am sources are cut up and sources are characterized individually with a low resolution gamma spectrometry system calibrated with an equal geometry with a secondary standard source of ^{241}Am or ^{226}Ra;
(v) All characterized sources from one head are put into a single plastic bag that is labelled with the reference number for the head and the individual source activities;
(vi) Sources are stored in Type B(U) packages for delivery to Amersham in the UK.

The following provides a brief overview of the conditioning facility (Fig. I–1):

(a) Radiation protection surveillance and system control area (cold area)①:
    (i) Operator entrance and exit;

(ii) Systems control (ventilation);
(iii) Radiation protection monitoring.

(b) Reception, storage, entrance and exit of radioactive materials and waste:
(i) Reception②;
(ii) Classification③;
(iii) Storage type A④.

(c) Conditioning and dismantling glovebox⑤.

(d) Characterization area⑥.

(e) B(U) type storage⑦.

(f) Facility maintenance (HEPA filter and pre-filter change, engine area with differential pressure monitoring devices, etc.)⑧.

The operational conditions at the facility have to fit the conditions of a class II radioactive facility, according to the Spanish regulations.

The glovebox (Fig. I–2) is a combined high integrity/air containment suite to provide clean air and easily decontaminated environment while protecting the operator from exposure to hazardous or toxic materials.

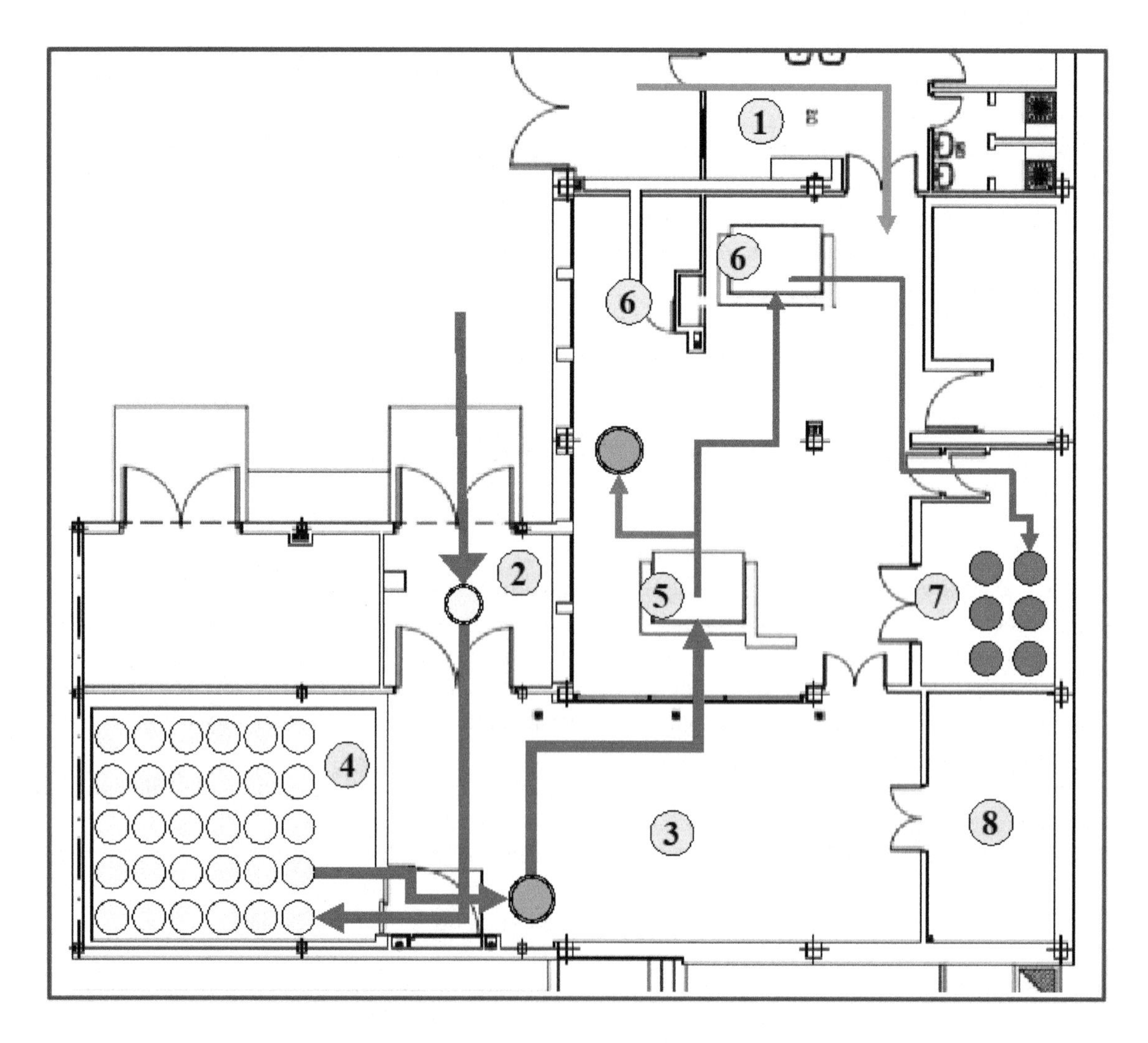

FIG. I–1. Overview of the conditioning facility.

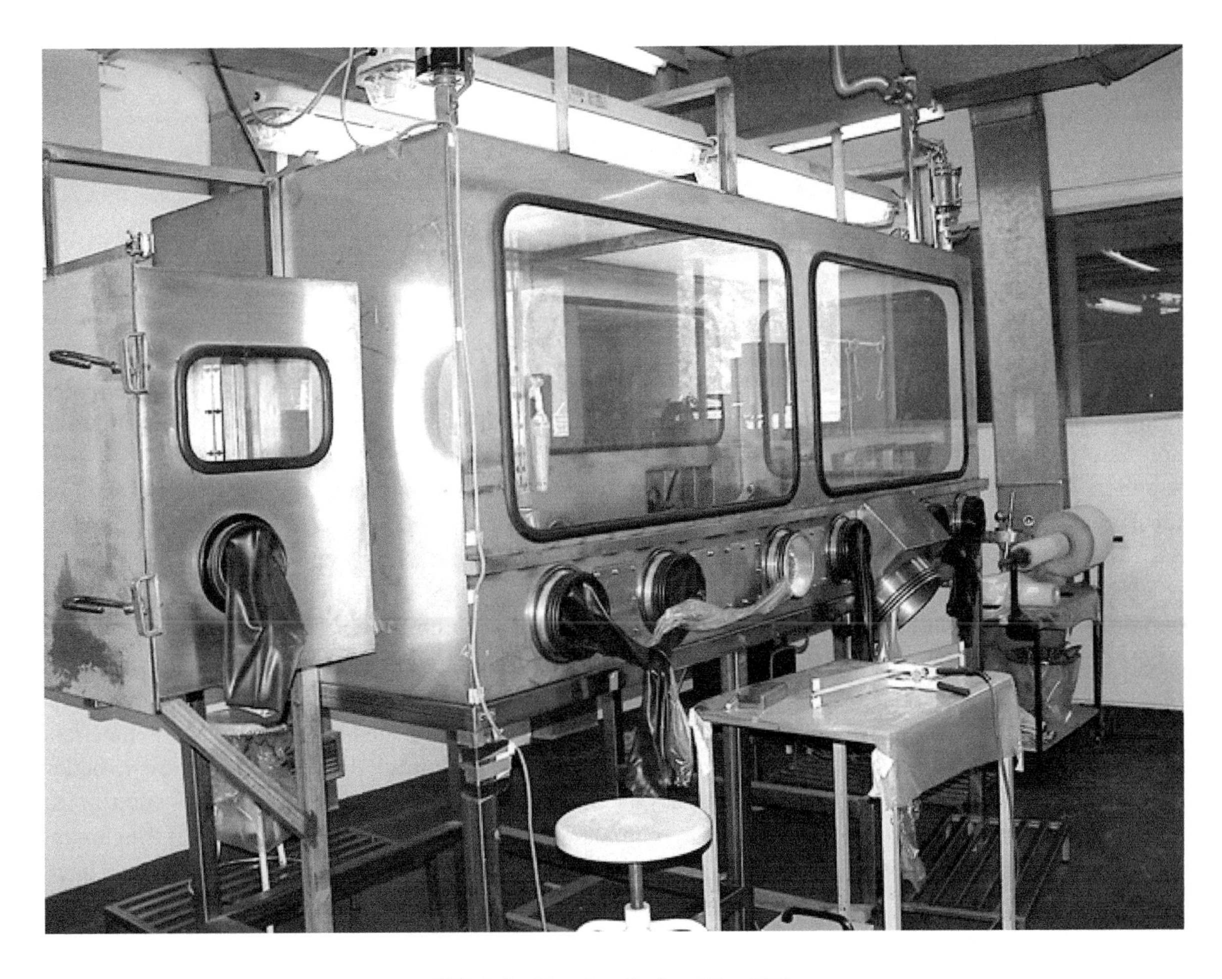

FIG. I–2. Glovebox for handling RLCs.

The beginning of the retrieval procedure starts with the location or advertisement of the owner by official communication according to a format that has to have the following data:

(a) Owner declaration:
  (i) Proper or legal representative of owner to sign.
(b) Location and accessibility to lighting conductors:
  (i) Address: city, street, etc.;
  (ii) Location: roof, cornice, etc.;
  (iii) Height of the conductor;
  (iv) Accessibility: window, trap door, staircase, etc.
(c) Data for the lighting conductor:
  (i) Brand;
  (ii) Model;
  (iii) Radionuclide.
(d) Owner data:
  (i) Entity/organization (if applicable);
  (ii) Name;
  (iii) Address;
  (iv) Telephone/fax;
  (v) Contact person (if applicable).
(e) Record keeping (see Fig. 4) starts with the declaration.

Annex II

EXAMPLE OF THE MANAGEMENT OF RLCs REMOVED FROM THE PUBLIC DOMAIN: SINGAPORE

II–1. SUMMARY

In Singapore, low activity or short lived radioactive sources are generally stored until the radioactive isotopes have decayed to below clearance levels before being disposed of at a landfill site. Such wastes are mainly generated by hospitals and research facilities. However, SRSs are to be returned to overseas suppliers at the end of their useful life. Such sources are mainly used in industrial applications such as level gauges, gamma irradiators, etc.

Through effective licensing controls, the authority ensures that SRSs imported into Singapore are re-exported to the country of origin when they are no longer in use. Nevertheless, over the years, the authority has accumulated a few hundred small disused SRSs (DSRSs). These DSRS were mainly in RLCs that could not be returned to their country of origin, since the manufacturing companies were no longer in operation. The RLCs were use in the early 1970s and are associated with radioactive materials such as ^{226}Ra and ^{241}Am. It was believed that the high charge density created by the radioactive materials would enhance the efficiency of the lightning conductor. However, studies indicate that RLCs are no better than conventional types of lightning conductors. Furthermore, in view of possible contamination risks, RLCs are no longer popular and, in fact, many companies have stopped their production. In Singapore, most of the RLCs have been removed and replaced by conventional types. Thus far, ~450 RLCs have been collected, with the sources removed and conditioned for storage.

II–2. ADMINISTRATIVE CONTROL OF RLCS

The control of RLCs began when the Radiation Protection Act was first implemented in 1973 to control the import, export, sale, transport, possession and use of radioactive materials and irradiating apparatus. Since then, RLCs have been identified as essential controlled items. Officers were deployed to gather information and locate potential RLC sites. More than 300 sites were located, and licences were issued to the owners. Such administrative controls were put in place to ensure that the inventory was properly managed and tracked. The types of RLCs that were used locally are outlined in Table II–1.

II–3. LONG TERM SOLUTION FOR RLCS

It was not until recently that the authority recognized the need to eradicate any potential security risk from any theft or terrorist act arising from the misuse of RLCs and began to look into a long term solution for the storage of RLCs. In 2004, the authority embarked on a project to encourage companies/owners to remove RLCs and assist them in the conditioning process. The companies that participated in this programme were responsible for engaging licensed contractors to remove RLCs and deliver them to the authority for eventual treatment. The companies were also required to pay a nominal disposal fee, their corresponding possession licences were terminated, and the companies were absolved of all subsequent liabilities pertaining to their RLCs. Although this programme was not mandatory, the response was successful and resulted in ~450 RLCs being collected from the public domain.

TABLE II–1. TYPES OF RLC INSTALLED IN SINGAPORE

Maker/manufacturer	Model	Isotope	Activity (mCi)
British Lightning Protection (BLPL)	P1	^{226}Ra	0.07
	P2		0.17
	P3		0.47
	P4		0.72
E.F. Australasia	S100	^{241}Am	2.14
	S150		3.21
	S200		4.28
	S500		9.62
Helita	AMH1	^{241}Am	0.20
	AMH2		0.30
	AMH3		0.45
	AMH4		0.60
	AMH5		0.75
Indelco	PAM1	^{226}Ra	0.07
	PAM2		0.17
Ionocaptor	FC3	^{241}Am	2.15
	FC4		2.60

II–4. IAEA ASSISTANCE

As part of the IAEA Technical Cooperation National Project Implementation Plan, Singapore sought IAEA assistance for a radioactive waste expert to advise on minimizing (reduction in volume) and conditioning of the accumulated RLCs for long term secure storage (see Figs II–1 and II-2). The waste conditioning project was carried out in two phases. The first phase involved reducing the waste volume through the separation of the DSRSs from other non-radioactive materials (e.g. casing of the lightning conductor), and this was carried out locally under the guidance of an IAEA expert.

The second phase, which involved conditioning and sealing of the DSRSs in drums for long term secure storage, was carried out jointly with the Korean Atomic Energy Research Institute (KAERI), who were engaged by the IAEA. This process involved encapsulation of the sources in steel containers and encasement of the steel containers in concrete drums.

In view of the success of the conditioning operation in Singapore, ~3000 pieces of sources, with a total of activity 600 mCi, were conditioned safely in stainless steel capsules that were placed in lead shielded containers and stored in concrete lined drums.

FIG. II–1. Removal of a radioactive source from a lightning conductor head.

FIG. II–1. Placing the lead container into a concrete lined drum during conditioning carried out by local experts and KAERI.

Annex III

EXAMPLE OF THE MANAGEMENT OF RLCs REMOVED FROM THE PUBLIC DOMAIN: CUBA

Radioactive lightning conductors (RLCs) have been used in Cuba for several decades; hundreds of these RLCs had been installed around the country, mainly in industrial settings. Figure III–1 shows some models of the RLCs most commonly used in the country.

III–1. REGULATORY CONSIDERATIONS

An inventory of radioactive lightning conductors installed in the country exists at the regulatory authority, the National Centre for Nuclear Safety (CNSN), since all the facilities have to be authorized for the installation and use of RLCs.

The use of RLCs does not comply with the justification principle established in the Basic Safety Standards [III–1]. Following international suggestions, the Ministry of Science, Technology and Environment (CITMA) approved Resolution 58/2003 [III–2], prohibiting the import and installation of new RLCs. The dismantling of RLCs that had been in use and replacement with conventional models over a 10 year period (before 2013) was also established in this regulation. Radioactive lightning conductors have to be transferred to the Centre for Radiation Protection and Hygiene (CPHR) for management as radioactive waste.

III–2. REMOVAL OF RADIOACTIVE LIGHTNING CONDUCTORS

The organizations involved in the removal of RLCs have to be qualified for this kind of work. A procedure describing the radiation protection requirements (administrative and technical) was developed and approved by the regulatory body. In most cases the operations are supervised directly by the regulators.

Radioactive lightning conductors were removed from facilities and replaced with conventional lightning conductors following established procedures. The RLCs were then placed in a container for

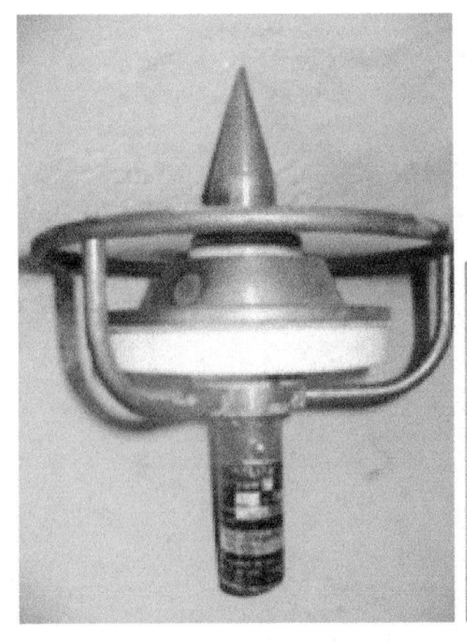

Helita (pellets/disks)

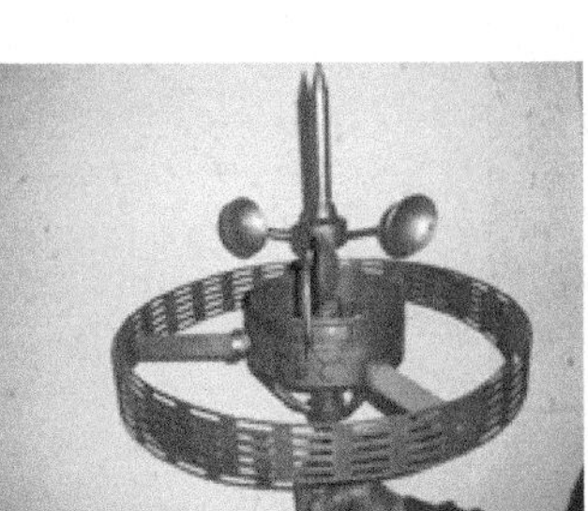

Nuclear Ibérica, Ionocaptor

Energía Fría

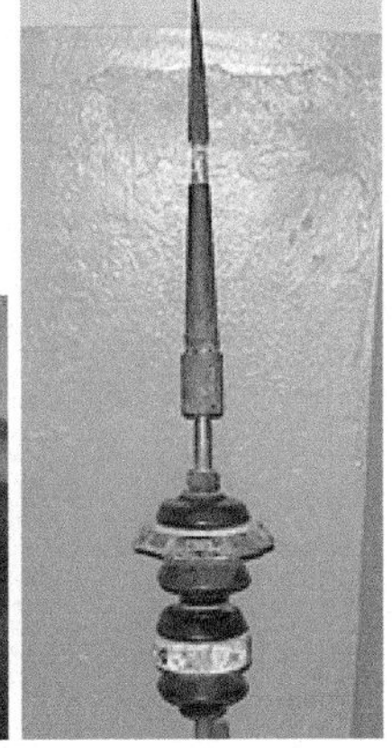

Duval Messien, Saref

FIG. III–1. The most common RLCs used in Cuba.

later collection by the CPHR for their further management. The CPHR is the responsible organization and is authorized by the CNSN for centralized management of radioactive waste in Cuba.

In some cases, RLCs have fallen because of strong winds during hurricanes. Measures established in the emergency plan of a radioactive facility are followed in these cases. The RLCs were recovered and inspected in order to verify whether all radioactive sources were in place. A detailed evaluation of the area for radioactive contamination and for the location of any lost sources was carried out. The RLCs were stored temporarily in the facility waiting for collection by the CPHR.

III–3. TRANSPORTATION OF RLCS TO THE WASTE MANAGEMENT FACILITIES, CHARACTERIZATION AND TEMPORARY STORAGE

Radioactive lightning conductors are transported to the waste management facilities by the CPHR, following the requirements established in the national transport regulations [III–3]. These RLCs are placed in plastic bags inside 200 L drums. Contamination and radiation levels are then measured and registered.

Upon arrival at the waste management facilities, RLCs are placed in interim storage. Each device is given an identification number and recorded in the Registry of Disused Radioactive Sources. Most RLCs used in Cuba contain ^{241}Am sources. Occasionally ^{226}Ra, ^{85}Kr and ^{14}C are also found. Some RLCs have to be characterized, as no information is available regarding the radionuclide contained and the activities of the radioactive sources. The radionuclide is identified using portable spectrometers (this was applied for sources containing gamma emitting radionuclides). The activity of the sources is estimated from the dose rate at a certain distance using the geometry of a point source.

III–4. DISMANTLING OF RLCS AND RECOVERY OF THE RADIOACTIVE SOURCES

Radioactive lightning conductors are dismantled in order to recover the radioactive sources for further conditioning and safe storage. A technical manual has been developed with specific instructions for dismantling each model of RLC stored in the facility. Radiation protection measures are an important issue when considering operations.

Contamination control is the principal concern when handling alpha emitting sources from lightning conductors (^{241}Am and ^{226}Ra). Adequate ventilation and filtration system have been installed in the working areas (a hood with a plastic wall tent and a mobile ventilation/filtration system (Fig. III–2)) to protect personnel and the environment from airborne contaminants, as loose contamination was expected.

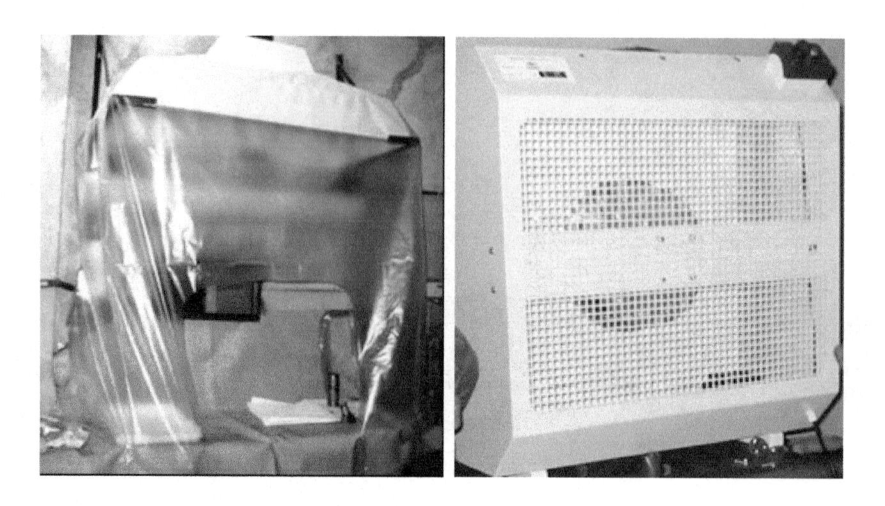

FIG. III–2. Hood with plastic wall tent and mobile ventilation/filtration system.

The benches and walls that may be exposed to contamination are covered with thick plastic film. A tray covered with absorbent paper is placed inside the hood and the lightning conductor head is placed there for dismantling operations. Appropriate personal protective equipment was used to prevent internal and external contamination of operators.

Stainless steel capsules were prepared in which to place the radioactive sources recovered from the lightning conductors. They were identified according to the record keeping system used in the facility. Only similar sources (same radionuclide, physical form, composition) would be placed in the same capsule.

Different containers were prepared for contaminated and non-contaminated materials generated during dismantling operations (Fig. III–3).

Radioactive lightning conductors are dismantled in the working area (inside the ventilation hood). First the dose rates at the surface of the sources are measured and contamination controls (using the wipe test) are carried out.

Detailed instructions were developed for dismantling each model of RLC stored in the facility. Figures III–4 and III–5 show examples of dismantling operations for the most common RLC used in Cuba: Helita and Ionocaptor.

FIG. III–3. Capsules in which to place radioactive sources recovered from RLCs, and contaminated and non-contaminated items generated during dismantling.

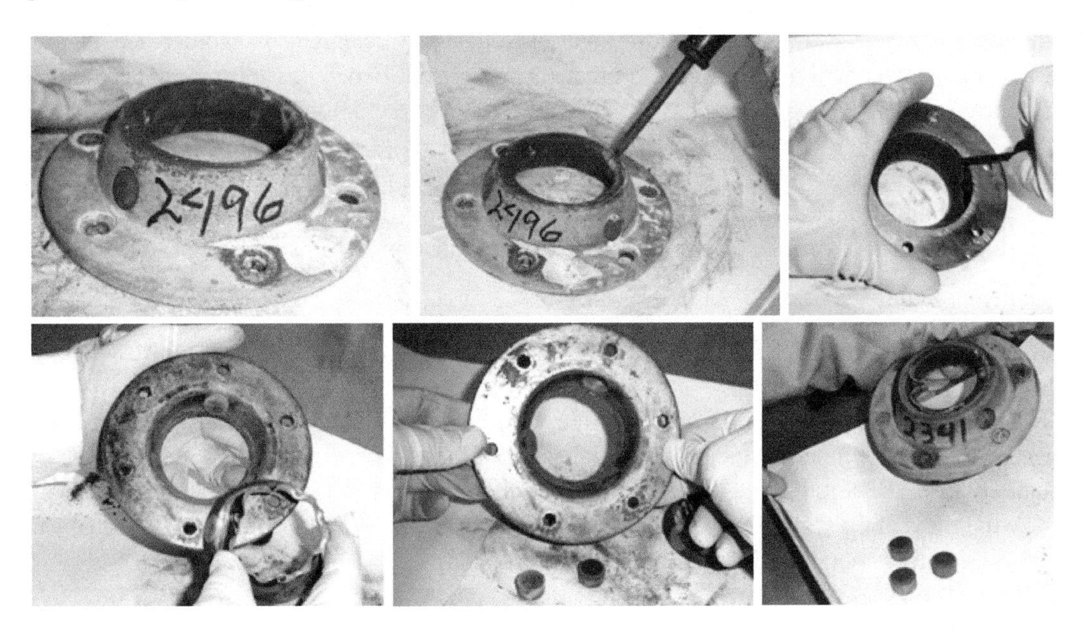

FIG. III–4. RLC, Helita: dismantling operations.

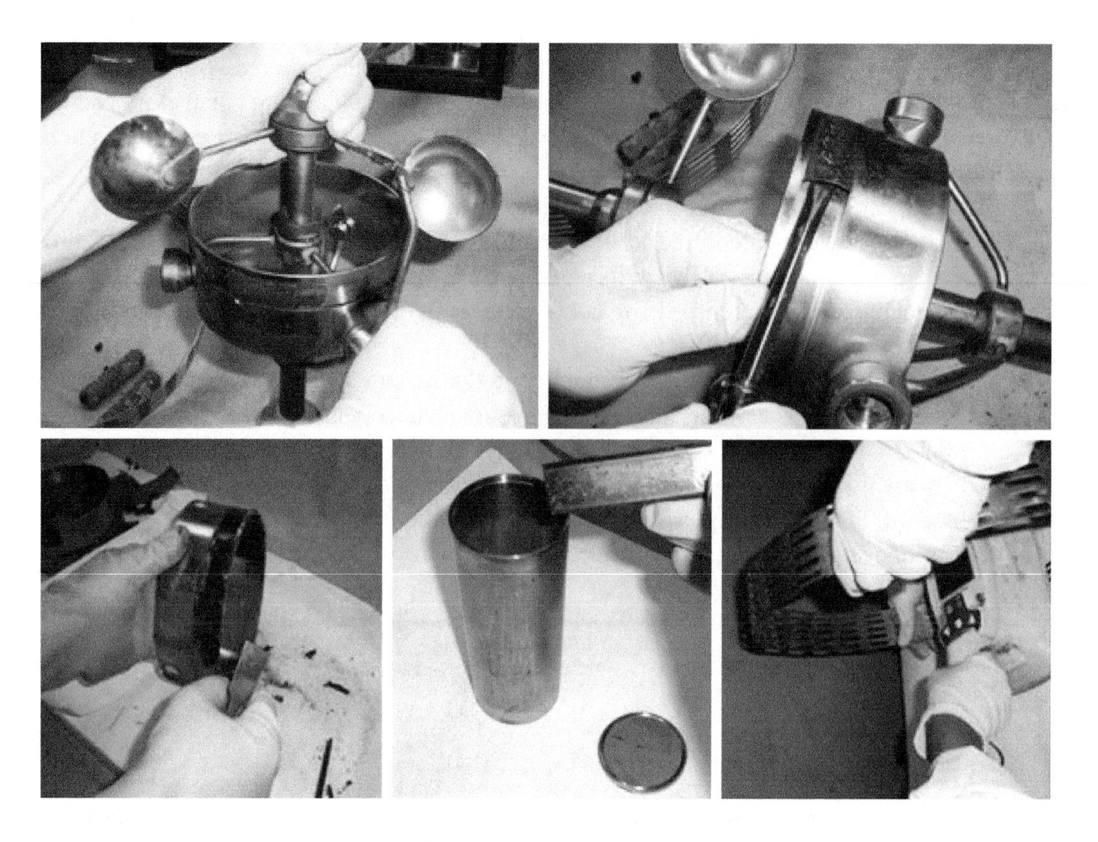

FIG. III–5. RLC, Ionocaptor: dismantling operations.

Recovered radioactive sources are placed in the corresponding capsule, depending on the radionuclide and physical characteristics. The records for dismantled RLCs are completed with information about the lightning conductor and the radioactive sources.

The total activity in the capsule is estimated from the dose rate measured at a certain distance from the sources using the geometry of a point source.

Once radioactive sources have been removed from the working area, the other components of the lightning conductor are measured to evaluate radioactive contamination (Fig. III–6). If radioactive contamination is detected, decontamination is carried out using detergent or another decontamination solution. If the contamination is fixed, contaminated parts are cut (when possible) to reduce volume and placed in heavy duty polyethylene bags, and then in the container previously prepared for solid radioactive waste for further storage. Non-contaminated items are managed as conventional wastes, verifying that there is no identification as radioactive material or the radioactive material symbol.

The working area is cleaned using absorbent paper and detergent or another decontamination solution (Fig. III–7). It is verified that no radioactive contamination remains in the working area.

III–5. CONDITIONING OF RADIOACTIVE SOURCES RECOVERED FROM RLCS

Radioactive sources recovered from RLCs are conditioned in stainless steel capsules. The number of sources in a capsule is limited by the physical dimensions of the sources, as the activities of these sources are relatively low. The capsules are sealed by welding, using a tungsten inert gas welding machine.

The tightness of the capsules is verified according to the ISO 9978 Standard [III–4]. Sealed capsules are stored inside lead containers in pre-cemented 2 L drums.

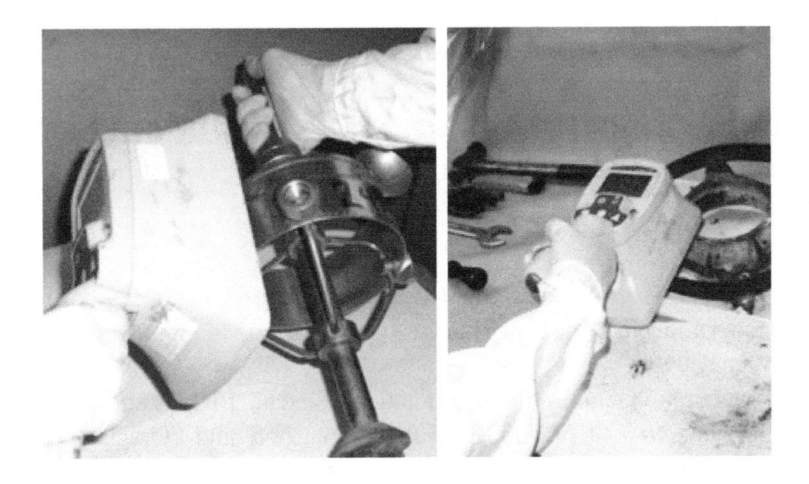

FIG. III–6. Measuring the radioactive lightning conductors after removal of the radioactive sources to evaluate radioactive contamination.

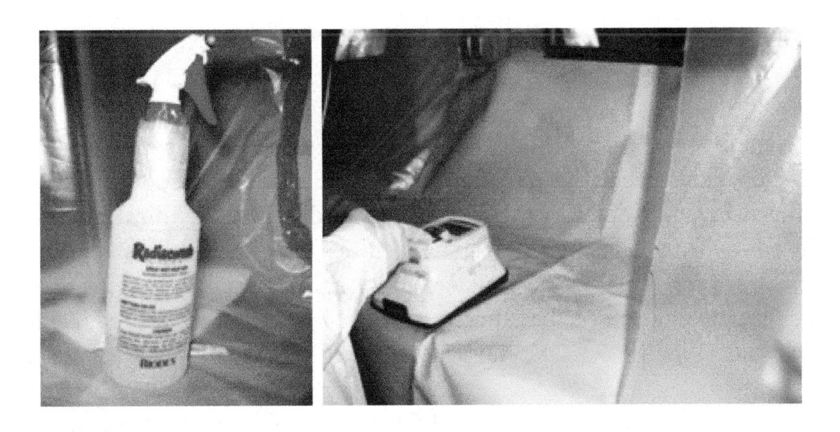

FIG. III–7. Cleaning and monitoring of the work area after dismantling operations.

III–6. CONCLUSIONS

Almost 200 RLCs have been removed from the facilities where they were installed and dismantled, with the radioactive sources being recovered and conditioned in stainless steel capsules for safe storage.

Special attention has been given to radiation safety issues; as alpha emitting sources are being handled and the risk of radioactive contamination exists.

Some RLCs are still installed in different facilities, as some financial limitations exist for replacing them with conventional models.

REFERENCES

[III–1] CITMA-MINSAP, Joint Resolution CITMA-MINSAP, Basic Safety Standards, Gaceta Oficial de la República de Cuba, Ordinary Edition, No. 1, La Habana (2002).

[III–2] CITMA, Resolution 58/2003, Gaceta Oficial de la República de Cuba, Ordinary Edition, No. 26, La Habana (2003).

[III–3] CITMA, Resolution 121/2000, Gaceta Oficial de la República de Cuba, La Habana (2003).

[III–4] INTERNATIONAL ORGANIZATION FOR STANDARDIZATION, Radiation Protection — Sealed Radioactive Sources — Leakage Test Methods, ISO 9978:1992, ISO, Geneva (1992).

Annex IV

EXAMPLE OF THE MANAGEMENT OF RLCs REMOVED FROM THE PUBLIC DOMAIN: PARAGUAY

Paraguay requested assistance from the IAEA to establish a national program for the management of radioactive lightning conductors (RLCs). This assistance was implemented through an expert mission, which took place in June 2010.

The national programme includes the identification of existing RLCs and updating of the national inventory; a plan for the removal of RLCs; removal, transport and storage of RLCs at the waste management facility; dismantling of RLCs, recovery and conditioning of the radioactive sources; procedures for all the operations; radiation safety and security measures; identification of all parties involved and specification of their responsibilities; as well as financial and human resources.

IV–1. INVENTORY OF RLCS

A national inventory of RLCs was not available at the regulatory body. The existing inventory only included the RLCs stored at the centralized storage facility at the National Commission of Atomic Energy (CNEA). The national inventory of RLCs still installed in different facilities around the country and stored at the site needs to be updated, although some facilities have been already registered.

Around 50 RLCs were stored at the CNEA facility and reported in the inventory. Most of them were Amerion and Gamatec models. A metallic drum with 11 RLC heads was included in the inventory. The drum was opened (Fig. IV–1) and the RLCs were individually characterized and registered with a new code.

The number of sources in each device varied from two to five and the dimensions of the sources (metallic foils) were also different (21 mm × 6.5 mm and 21 mm × 11 mm). This information was also updated in the registry. After RLCs were removed from the drum, contamination control was carried out. The drum had to be decontaminated as some radioactive contamination was detected.

IV–2. NATIONAL PLAN FOR REMOVAL AND MANAGEMENT OF RLCS

Necessary information was provided to counterparts during the mission, including international experience and guidance, to prepare a national plan for removal of the installed RLCs and safe management. Elements to be considered in the national plan are described in detail: radiation protection considerations (justification and optimization principles); regulatory support; updating of the inventory

FIG. IV–1. RLC heads stored in a metallic drum.

and location of RLCs still installed; removal of RLCs and replacement with conventional models; authorization of operations. Once dismantled, the RLCs need to be transferred to the waste management facilities at CNEA for further management as radioactive waste.

IV–3. REMOVAL OF RLCS AND TRANSPORT TO THE WASTE MANAGEMENT FACILITY

Removal of RLCs will always be performed by qualified and authorized personnel. General instructions and radiation safety measures for removal of RLCs have been provided. Other conventional safety measures currently in force in Paraguay are also to be followed, mainly measures related to working at heights.

The transport of removed RLCs to the CNEA facilities has to be carried out following the national regulations. Decree 10754/2000, National Safety Regulations for Protection against Ionizing Radiations and for the Safety of Radioactive Sources, establishes that the transport of radioactive materials is to comply with the requirements established in the IAEA Transport Regulations.

IV–4. DISMANTLING RLCS AND RECOVERING THEIR RADIOACTIVE SOURCES

Dismantling operations are carried out at the CNEA facilities by trained and authorized personnel. A procedure was developed, and first operations were carried out during the expert mission, with the supervision of IAEA experts.

The working areas were prepared at the laboratory, and they were covered with plastic film to avoid the possible dispersion of radioactive contamination (Fig. IV–2).

The working place PT-01 is used to receive the RLCs from the storage facility (Fig. IV–3). Radiation measurements and contamination controls are carried out in this area. The RLCs are dismantled in working place PT-01 up to the source holder level (depending on the RLC model).

The components that do not contain the radioactive sources are transferred to the working place PT-04 for contamination controls (Fig. IV–4). Contaminated items are stored as radioactive waste. Non-contaminated parts can be managed as conventional waste.

The radioactive sources are removed from their source holders in the working place PT-02. The sources are placed individually in plastic bags and transferred to the spectrometric laboratory for characterization (Fig. IV–5). The radionuclide of the source is identified, and the activity is estimated.

The other components (not directly associated with the radioactive sources) are transferred to PT-04 for contamination controls and a decision about further management (as conventional or radioactive waste, Fig. IV–6). Decontamination of radioactive contaminated items needs to be evaluated, taking into consideration the minimization of radioactive waste generation.

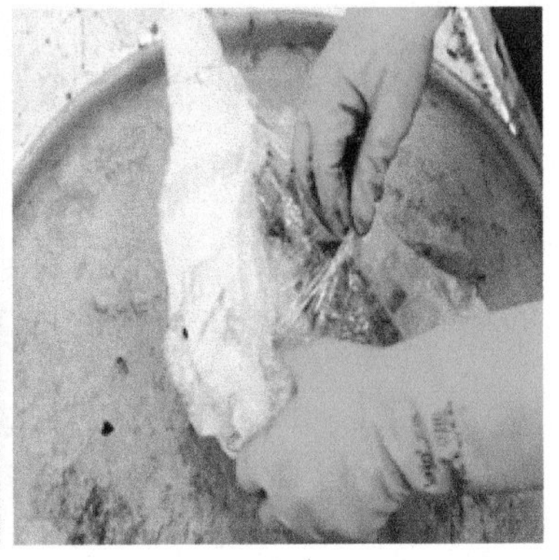

FIG. IV–2. Preparing the working areas for dismantling operations.

FIG. IV–3. Working place PT-01

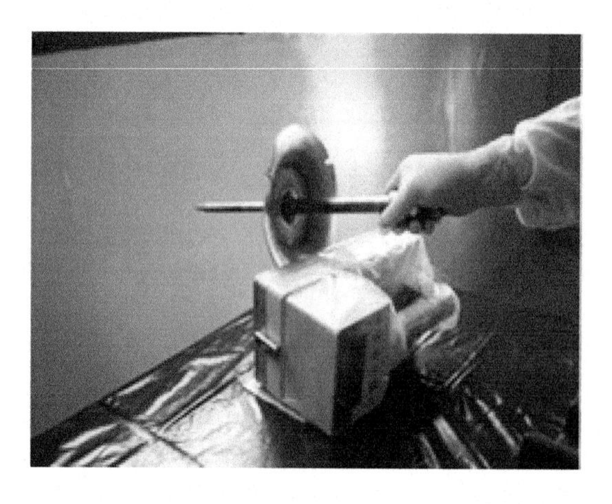

FIG. IV–4. Working place PT-04, contamination control.

FIG. IV–5. Characterization of the radioactive sources recovered from the RLCs.

IV–5. CONDITIONING OF RADIOACTIVE SOURCES RECOVERED FROM THE RLCS

Radioactive sources recovered from RLCs are conditioned in retrievable form, considering that no final disposal solution is available in the country. Conditioning operations are carried out in workplace PT-03.

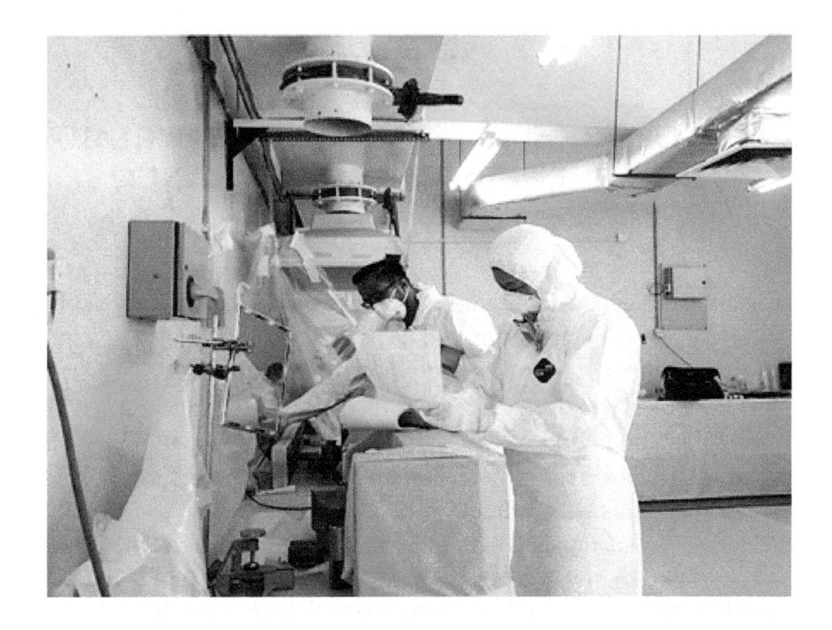

FIG. IV–6. Non-contaminated items generated from dismantling the RLCs.

FIG. IV–7. Conditioning of radioactive sources in a stainless steel capsule.

Sources are taken out from the plastic bags and placed in stainless steel capsules for conditioning. Radiation levels at the surface of each source and the capsule are measured and registered (Fig. IV–7). The capsule is closed with a tightly screwed on lid.

After operations are finished, the working area is measured for radioactive contamination.

The stainless steel capsule with the sources is placed in a container and transferred to the storage facility.

IV–6. CONCLUDING REMARKS

The methodology for the management of RLCs has been implemented in Paraguay, including dismantling RLCs and recovery and conditioning of radioactive sources.

A national plan for updating the national inventory of existing RLCs and their removal and safety management will be developed and implemented.

Annex V

EXAMPLE OF THE MANAGEMENT OF RLCs REMOVED FROM THE PUBLIC DOMAIN: FRANCE

In France, the National Agency for Radioactive Waste Management (ANDRA) is in charge of managing all radium bearing items by collecting and packaging them, especially the radioactive lightning conductors (RLCs) collected all over the country by dismantling companies.

According to an Order of 11 October 1983, it has been forbidden to use any radioactive element in the manufacturing, marketing and import of RLCs in France since 1 January 1987. However, this Order does not require the removal of RLC in the public domain. Around 500 RLCs are removed each year, one third with americium and two thirds with radium.

The heads of removed RLCs are placed in drums and transferred to ANDRA for storage. The heads are dismantled into separate radioactive sources from other components.

The average activity of RLC sources is 22 MBq for americium and 48 MBq for radium. No other radionuclides are in RLCs in France.

The best estimate is that 40 000 RLCs are still in place today. At the current removal rate, it will take 80 years to remove all the RLCs from the public domain in France.

Because RLCs were installed before the regulation on management of radioactive sources, there was neither a database nor an inventory of their locations.

A storage facility that was set up by ANDRA in October 2012 is able to receive all 40 000 RLCs.

The French Safety Authority has prepared a new order to impose a ban on the use of all RLCs.

Annex VI

EXAMPLE OF THE MANAGEMENT OF RLCs REMOVED FROM THE PUBLIC DOMAIN: MALAYSIA

VI–1. INTRODUCTION

The Malaysian Nuclear Agency, through its operating unit on waste management, the Waste Technology Development Centre, receives and collects lightning conductors for further management according to the safe and secure management of radioactive waste. These lightning conductors come in the form of the whole complete rod or the head part of it. The handling, packaging and transport of the lightning conductors and their parts will abide by the Radiation Protection (Transport) Regulations 1989 and the Atomic Energy Licensing (Radioactive Waste Management) Regulations 2011. As of 2016, the inventory of the lightning conductors stored at the national interim storage facility stood at 149 units. It was then decided that the radioactive sources in these lightning conductors were to be retrieved, encapsulated and ultimately disposed of in the planned borehole disposal facility.

To date, the Government of Malaysia is continuing its effort to motivate the removal of all remaining lightning conductors containing radioactive material from use while the new import and installation of RLCs have been terminated. The regulatory body, the Atomic Energy Licensing Board (AELB), estimates that approximately 500 old lightning conductors containing radioactive sources are still in the public domain and currently installed on buildings across the country.

VI–2. MANAGEMENT OF THE RADIOACTIVE LIGHTNING CONDUCTORS

The RLCs, once removed from their installations for the purpose of disposal, are treated as low level radioactive waste with potential contamination. The removal from the installation is carried out by competent persons, either Nuclear Malaysia or sub-contractors licensed by the AELB. Some of the rods have tags depicting the radioactive logo, and describing the radioactive element and activity and the official correspondence (Fig. VI–1).

The RLCs are sent to the Malaysian Nuclear Agency for further action. Most of the RLC rods are cut off on-site and the heads are then wrapped properly to prevent cross-contamination. However, receiving the complete RLC set with its long rod is also a common situation (see Fig. VI–2). A record of the received RLC is logged, and the information for the RLC is then incorporated into the national waste inventory list. All RLCs are stored at the interim storage facility until further management steps are decided upon and carried out.

VI–3. TYPES AND MODELS OF RLCS IN MALAYSIA

Table VI–1 summarizes the types and models of lightning conductors received and stored at the Malaysian Nuclear Agency's interim storage facility.

FIG. VI–1. An example of a tag affixed to the rod of an RLC. Courtesy of the Malaysian Nuclear Agency.

  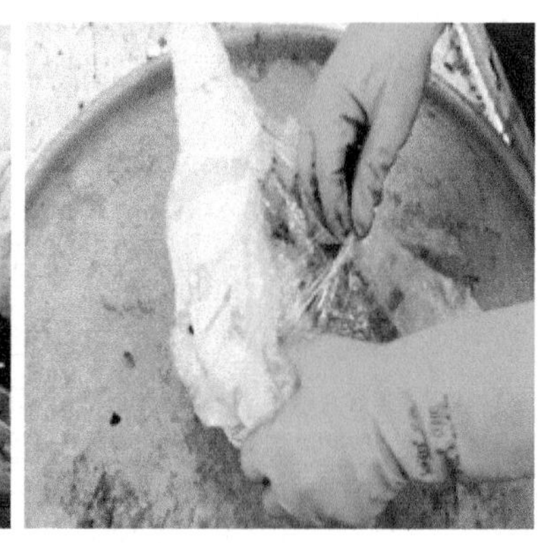

FIG. VI–2. RLCs received in the form of long rods or heads only. Courtesy of the Malaysian Nuclear Agency.

VI–4. RADIATION PROTECTION DURING THE DISMANTLING AND SOURCE RETRIEVAL OF THE RADIOACTIVE SOURCES

The campaign to dismantle lightning conductors with subsequent retrieval and encapsulation of the disused radioactive sources from these conductors began in December 2017 via an IAEA Expert Mission under the RAS9085 project.

All lighting conductors were being treated as contaminated objects and prone to disseminate surface contamination due to the degraded condition of the rods under the effects of weathering and corrosion. As such, the highest degree of contamination control was undertaken in order to minimize the spread of contamination from them. The following radiation protection measures were outlined and implemented during the exercise.

(a) Personnel:

   (i) Full PPE is worn by personnel during the entire dismantling, source retrieval and encapsulation process;
   (ii) Clear designation of task is given to all personnel. Each group comprises four team members (two persons for dismantling, one person for recording and one person for dose measurement/ contamination survey).

TABLE VI–1. DISUSED RADIOACTIVE LIGHTNING CONDUCTORS RECEIVED BY THE MALAYSIAN NUCLEAR AGENCY

Model	Type (activity)	Example of the RLC	Source information
EF 33 Australasia (Australia)		 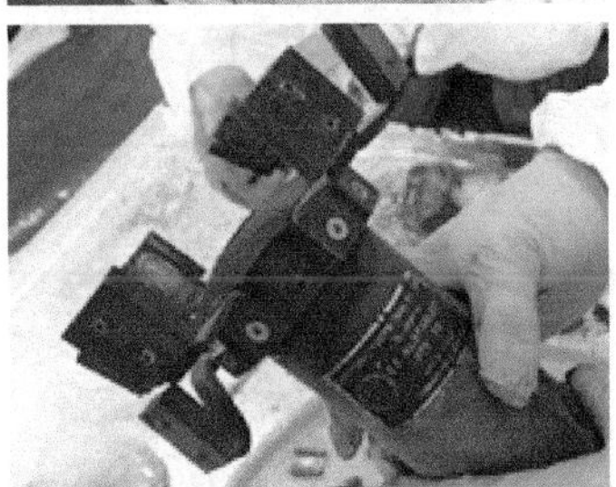  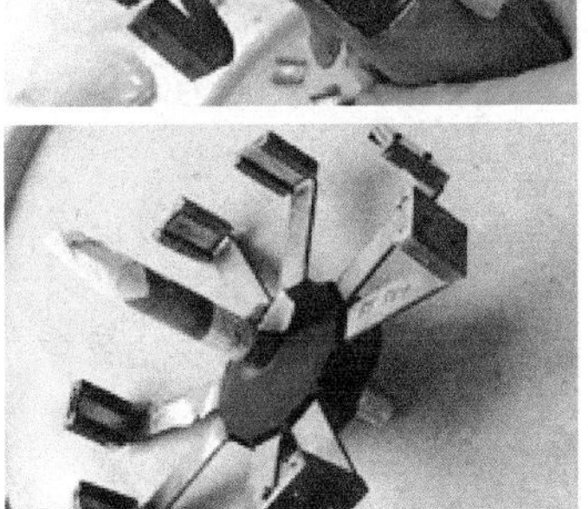	Source can be removed using an Allen key. If this is not possible, the metal slot containing the radioactive plate is cut off from the head 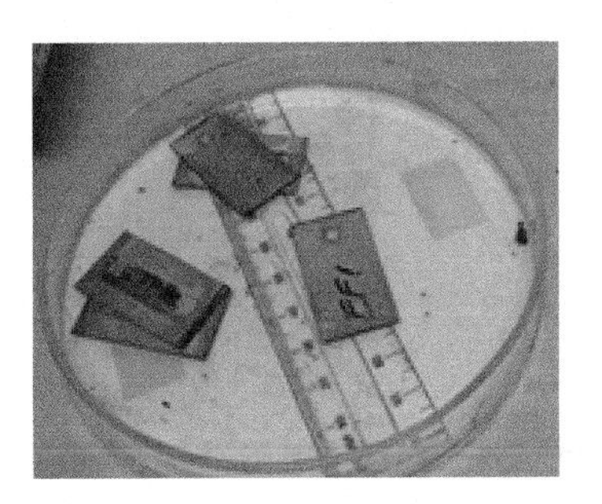
Helita (made 1979–1981)	AMH 5 (0.75 mCi) AMH 4 (0.6 mCi), AMH 3 (0.45 mCi)	 	^{241}Am Source type SAC 2 

TABLE VI–1. DISUSED RADIOACTIVE LIGHTNING CONDUCTORS RECEIVED BY THE MALAYSIAN NUCLEAR AGENCY (cont.)

Model	Type (activity)	Example of the RLC	Source information
Preventor	Type P1, type P2		^{241}Am 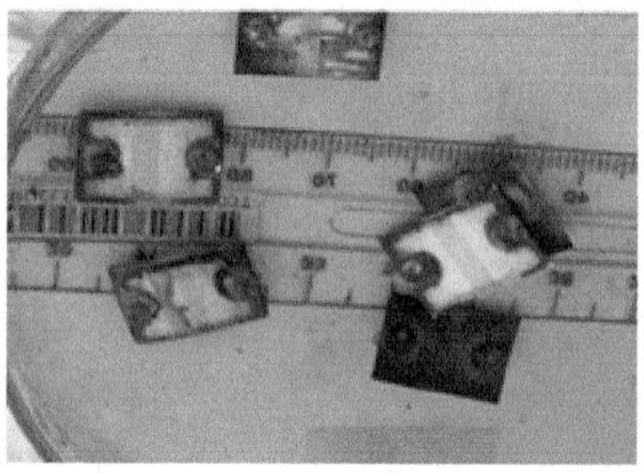
	PAM 3		^{241}Am 
Parasphere (France)		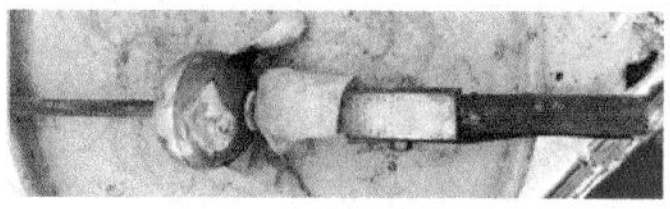	^{226}Ra 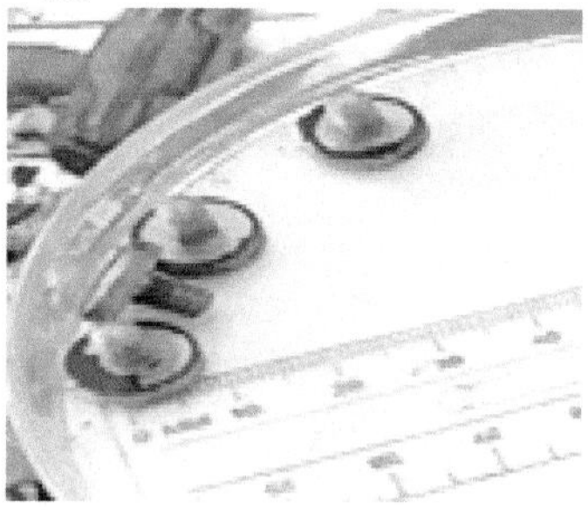 Sources are missing from the original slots in almost all heads
Saint Elme			^{241}Am 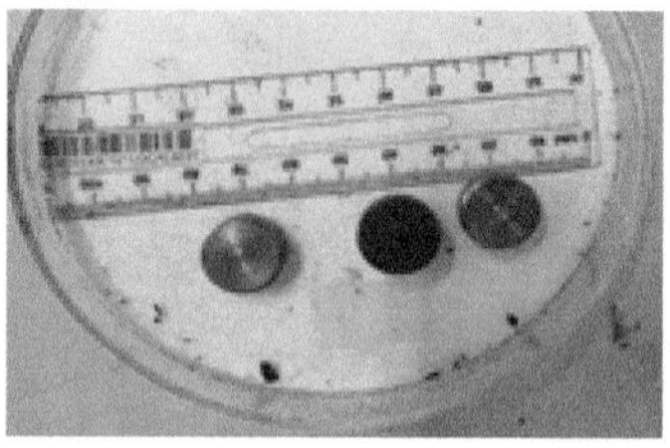

(b) Work area:

(i) The entire room, including the walls and floors, are protected with a thick layer of plastics;
(ii) Other items/tools in the room that are not involved and used in the process are either removed or covered;
(iii) There is clear demarcation of the work area to avoid cross-contamination by personnel. Areas are assigned as either clean or dirty zones. For example, the personnel responsible for

dismantling work in the dirty zone and are prohibited from leaving the zone or stepping into the clean zone until clearance is given;

(iv) Dismantling is carried out on a metal tray in order to collect any debris, metal pieces and rust (Fig. VI–3). The tray is also used to collect sources resulting from removal.

(c) Waste management:

(i) All metal pieces generated from the dismantling are surveyed and contamination levels are checked using contamination meters (Fig. VI–4);
(ii) Two drums are prepared in the work area, one for non-contaminated waste and one for contaminated waste;
(iii) The metal tray used in the dismantling work is cleaned and surveyed occasionally in-between dismantling in order to control contamination and minimize radioactive waste generation;
(iv) All tools use in the work are wrapped with plastic and tape as a protective layer against contamination. The tools are cleared for reuse after thorough cleaning and decontamination.

VI–5. ENCAPSULATION OF THE DSRSS

DSRSs from the RLCs are encapsulated in stainless steel capsules (160 mm × 66 mm). Six hundred and forty-seven DSRSs were retrieved from the 150 RLC units that were dismantled, comprising 417 ^{226}Ra sources and 230 ^{241}Am sources. These sources are contained in only four capsules, like the one shown in Fig. VI–5.

VI–6. DATA RECORDING

It is important to note that, unlike radioactive gauges, some of the sources from the RLCs are not marked to indicate their radioactive element and its activity. Some of the RLCs have this information on the metal plate label, but there are occasions when the metal plate has become totally unreadable. It

FIG. VI–3. A worker in the 'dismantling zone', marked in blue, using a hammer against a head in an attempt to remove the source. Courtesy of the Malaysian Nuclear Agency.

is therefore important to be equipped with an identifier to help determine and confirm the radioactive element when conditioning the RLCs.

Each source that has been retrieved is measured for its surface dose rate and the dose rate at 25 cm. The set-up to perform dose measurement is depicted in Fig. VI–6. To minimize background interference, the set-up is arranged behind the shielding. This dose rate reading and other details of the source and the RLC are recorded in a designated form that is later transferred into the electronic copy. It is important to keep both the hardcopy and softcopy records to retain them for future reference.

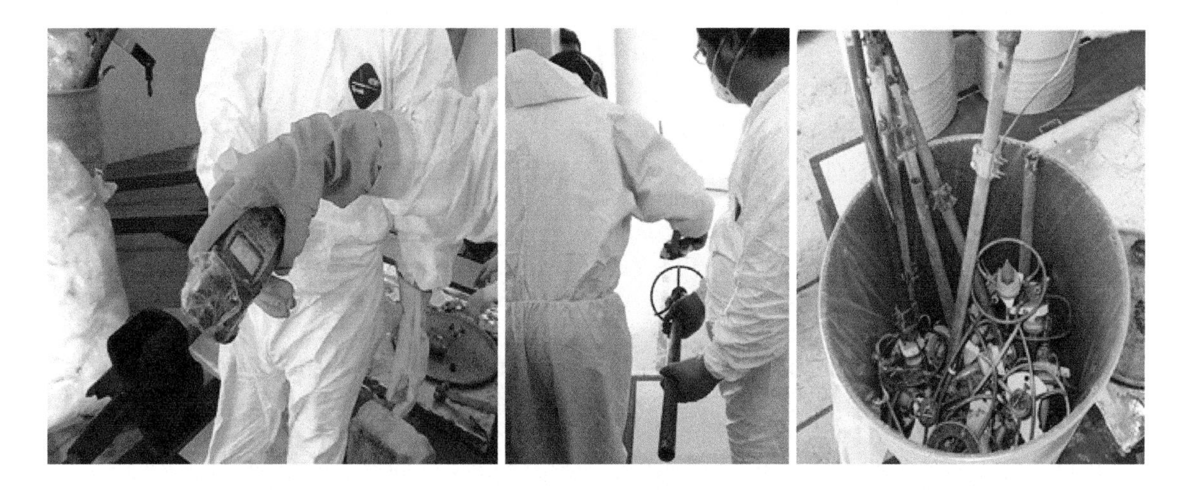

FIG. VI–4. Contamination survey using a RadEye B20 multipurpose survey meter. With proper contamination control measures, more than 60% of the metal pieces are categorized as non-contaminated waste. Courtesy of the Malaysian Nuclear Agency.

FIG. VI–5. DSRSs retrieved from RLCs were encapsulated in stainless steel capsules. Courtesy of the Malaysian Nuclear Agency.

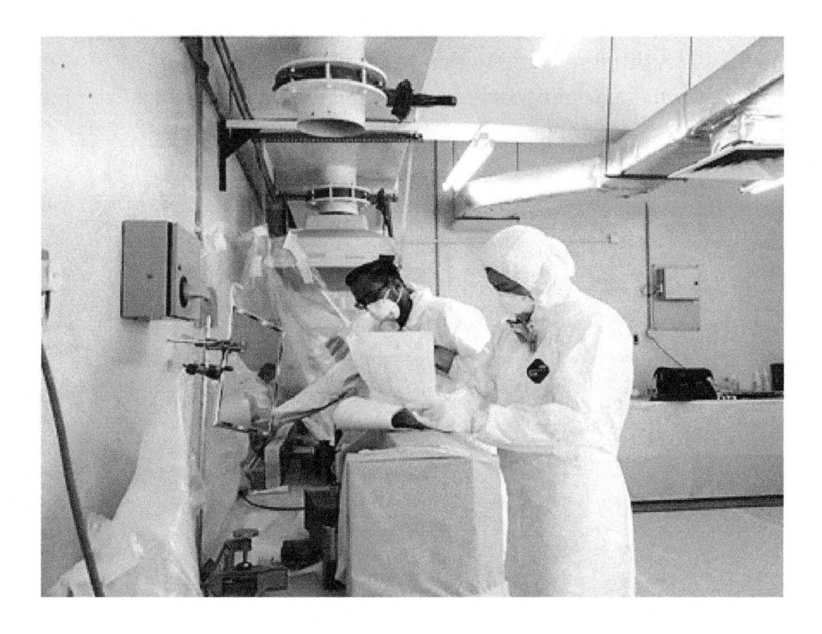

FIG. VI–6. Set-up for measuring the dose rate for each source from RLCs. Courtesy of the Malaysian Nuclear Agency.

Annex VII

EXAMPLE OF THE MANAGEMENT OF RLCs REMOVED FROM THE PUBLIC DOMAIN: MALTA

^{241}Am lightning conductors have been in use in Malta for some years, and are largely found on government buildings, churches and hotels.

The Commission for the Protection from Ionizing and Non-Ionizing Radiation (Commission), which is the regulatory authority for Malta, is aware of the issues presented by RLCs and is in the process of having them removed and placed into safe storage.

Under the latest Maltese regulations, RLCs have now become a non-justified use of ionizing radiation and, as such, they have to be removed and stored in Malta's centralized storage facility.

Due to the fact that the RLCs were installed many years ago, some of the building owners are unaware that the buildings have lightning conductors at all, let alone that these contain a radioactive material. Many are genuinely surprised when confronted with the issue and the fact that management of the RLCs is compulsory.

Unfortunately, there have been instances in the past when RLCs have been removed by their owners and improperly stored or discarded on the premises.

The current programme for removal requires that the owner, or legal person, notifies the Commission about the RLC on their premises. The Commission then orders the RLCs to be taken down and moved to the national radioactive waste store by an approved person.

The two RLC models that are encountered in Malta are shown in Fig. VII–1, both of which have ^{241}Am as the radioactive component.

During the removal of an RLC from a building a protective foam is inserted into the spike (Fig. VII–2). This is to prevent injury to workers and puncturing of the bag in which the RLC is placed to prevent any potential spread of contamination.

Once taken from the site, the RLCs are conditioned by cutting off the surrounding plate using steel shears (RLCs containing ^{241}Am foil sources) or simply dismantling the whole unit and removing the pellets (RLCs containing ^{241}Am ceramic sources) (Fig. VII–3).

Once conditioned, the sources are counted, inventoried and placed inside a stainless steel capsule, which then goes into a Type A drum residing within the storage facility.

It is pertinent to mention that Malta received much invaluable assistance from the IAEA in the form of an expert mission, training and equipment.

FIG. VII–1. The two models of radioactive lightning conductors encountered in Malta. Courtesy of the Radiation Protection Commission, Malta.

a)

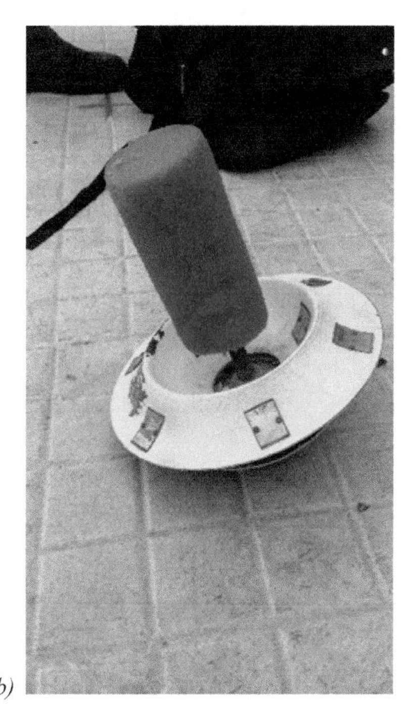

b)

FIG. VII–2. (a) Removal of an RLC from a building. (b) Note the protective foam on the spike. Courtesy of the Radiation Protection Commission, Malta.

FIG. VII–3. RLC assemblies after the radioactive sources have been removed. Courtesy of the Radiation Protection Commission, Malta.

Due to Malta's geographical size and landscape, RLCs are relatively easily identified, although it is wise to keep in mind that some might still exist without the Commission's knowledge, and most importantly, that not all owners are aware of their legal obligations.

As an example, one Sunday evening a Commission staff member went for a walk and passed down a quiet road behind a local hotel; an object in one of the adjoining fields caught the eye of this person and, on closer inspection, it was discovered that the object was an abandoned RLC (Fig. VII–4). From the road the object looked similar to a boat propeller, which would have been reasonable, given that the hotel sits on a beach.

The following day (Monday), Commission inspectors went to the area and confirmed that the object was an RLC with ^{241}Am foils still attached (Fig. VII–5).

The RLC was removed by an approved person under the Commission inspectors' supervision and stored safely and securely at the hotel premises until it was transported to the Malta Central Storage facility in December 2018. During removal, it was noted that several of the ^{241}Am foils displayed deterioration and weathering. Once the RLC was removed, the soil underneath was sampled and several pieces of active foil (6 mm^2 or less) were found. These were sealed and stored within a plastic bag. Some of the soil was also gathered and sealed in a plastic bag due to localized contamination.

All the material was placed within a heavy duty plastic bag, which in turn was placed in a container. Sand was used to cover the entire bag, whilst still making it accessible for future retrieval. Finally, the container was closed with a lid and a numbered seal was attached to it. Figure VII–6 shows the final package.

Upon licensing of the central storage facility in Malta, the hotel was asked to immediately arrange for the RLC to be removed and transferred to the storage facility for conditioning. The conditioning area for handling disused radioactive sources is shown in Fig. VII–7.

FIG. VII–4. An abandoned RLC. Courtesy of the Radiation Protection Commission, Malta.

FIG. VII–5. An abandoned and damaged RLC with ^{241}Am sources. Courtesy of the Radiation Protection Commission, Malta.

FIG. VII–6. Final package with the recovered RLC and ^{241}Am foils. Courtesy of the Radiation Protection Commission, Malta.

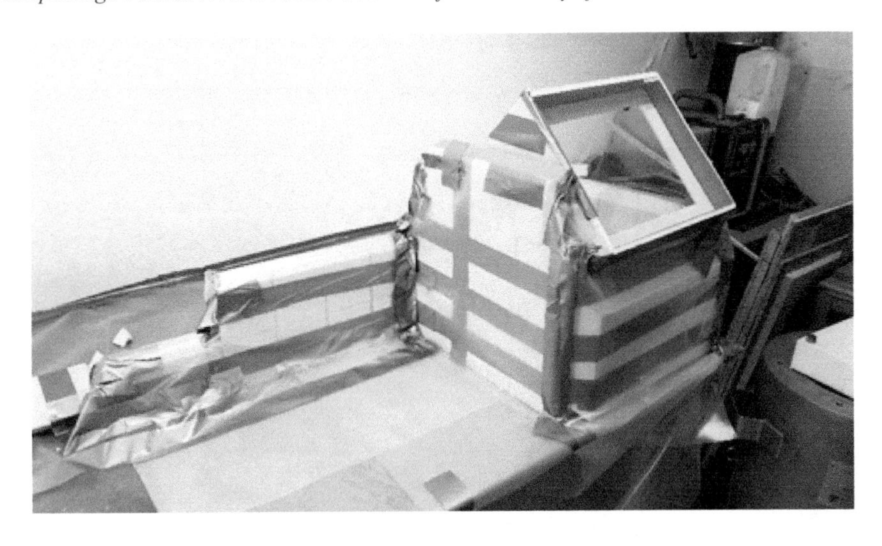

FIG. VII–7. Source conditioning area at the waste storage facility Courtesy of the Radiation Protection Commission, Malta.

ABBREVIATIONS

ALARA	as low as reasonably achievable
ANDRA	National Radioactive Waste Management Agency, France
CPHR	Radiation Protection and Hygiene Center, Cuba
DSRS	disused sealed radioactive source
IDB	information database
ISO	International Organization for Standardization
ITN	Nuclear and Technological Institute, Portugal
mSv	millisievert
PPE	personal protective equipment
RLC	radioactive lightning conductor
SRS	sealed radioactive source
TAEK	Turkish Atomic Energy Authority

CONTRIBUTORS TO DRAFTING AND REVIEW

Benitez-Navarro, J.C.	International Atomic Energy Agency
Bennett, D.	International Atomic Energy Agency
Cremona, J.	Radiation Protection Commission, Malta
Csullog, G.W.	Consultant, Canada
Deconinck, J.M.	Consultant, Belgium
Demetriades, P.	Consultant, Cyprus
Djurovic, T.	Ministry of Sustainable Development and Tourism, Montenegro
Durand, A.	ANDRA, France
Friedrich, V.	International Atomic Energy Agency
Heard, R.	International Atomic Energy Agency
Hoo, W.T.	National Environment Agency, Singapore
Hordijk, L	Necsa, South Africa
Hossain, S.	Consultant, Austria
Kinker, M.	International Atomic Energy Agency
Marumo, J.T.	IPEN, Brazil
Neubauer, J.	Consultant, Austria
Novakovic, M.	Ekotech, Croatia
Piña Lucas, G.	Center for Energy, Environmental and Technological Research, Spain
Salgado, M.	CPHR, Cuba
Sakkas, D.	DLI, Cyprus
Stewart, W.	International Atomic Energy Agency
Vicente, R.	IPEN, Brazil
Zakaria, N.	Malaysian Nuclear Agency, Malaysia

Consultants Meetings

Vienna, Austria: 12–16 December 2011; 3–7 June 2013

7–11 April 2014; 11–15 May 2015

Structure of the IAEA Nuclear Energy Series*

Nuclear Energy Basic Principles
NE-BP

Nuclear Energy General Objectives
NG-O

1. Management Systems
NG-G-1.#
NG-T-1.#

2. Human Resources
NG-G-2.#
NG-T-2.#

3. Nuclear Infrastructure and Planning
NG-G-3.#
NG-T-3.#

4. Economics and Energy System Analysis
NG-G-4.#
NG-T-4.#

5. Stakeholder Involvement
NG-G-5.#
NG-T-5.#

6. Knowledge Management
NG-G-6.#
NG-T-6.#

Nuclear Reactor Objectives**
NR-O

1. Technology Development
NR-G-1.#
NR-T-1.#

2. Design, Construction and Commissioning of Nuclear Power Plants
NR-G-2.#
NR-T-2.#

3. Operation of Nuclear Power Plants
NR-G-3.#
NR-T-3.#

4. Non Electrical Applications
NR-G-4.#
NR-T-4.#

5. Research Reactors
NR-G-5.#
NR-T-5.#

Nuclear Fuel Cycle Objectives
NF-O

1. Exploration and Production of Raw Materials for Nuclear Energy
NF-G-1.#
NF-T-1.#

2. Fuel Engineering and Performance
NF-G-2.#
NF-T-2.#

3. Spent Fuel Management
NF-G-3.#
NF-T-3.#

4. Fuel Cycle Options
NF-G-4.#
NF-T-4.#

5. Nuclear Fuel Cycle Facilities
NF-G-5.#
NF-T-5.#

Radioactive Waste Management and Decommissioning Objectives
NW-O

1. Radioactive Waste Management
NW-G-1.#
NW-T-1.#

2. Decommissioning of Nuclear Facilities
NW-G-2.#
NW-T-2.#

3. Environmental Remediation
NW-G-3.#
NW-T-3.#

(*) as of 1 January 2020
(**) Formerly 'Nuclear Power' (NP)

Key

BP:	Basic Principles
O:	Objectives
G:	Guides and Methodologies
T:	Technical Reports
Nos 1–6:	Topic designations
#:	Guide or Report number

Examples

NG-G-3.1:	Nuclear Energy General (**NG**), Guides and Methodologies (**G**), Nuclear Infrastructure and Planning (topic **3**), **#1**
NR-T-5.4:	Nuclear Reactors (**NR**), Technical Report (**T**), Research Reactors (topic **5**), **#4**
NF-T-3.6:	Nuclear Fuel (**NF**), Technical Report (**T**), Spent Fuel Management (topic **3**), **#6**
NW-G-1.1:	Radioactive Waste Management and Decommissioning (**NW**), Guides and Methodologies (**G**), Radioactive Waste Management (topic **1**) **#1**

No. 26

ORDERING LOCALLY

IAEA priced publications may be purchased from the sources listed below or from major local booksellers.

Orders for unpriced publications should be made directly to the IAEA. The contact details are given at the end of this list.

NORTH AMERICA

Bernan / Rowman & Littlefield

15250 NBN Way, Blue Ridge Summit, PA 17214, USA
Telephone: +1 800 462 6420 • Fax: +1 800 338 4550
Email: orders@rowman.com • Web site: www.rowman.com/bernan

REST OF WORLD

Please contact your preferred local supplier, or our lead distributor:

Eurospan Group

Gray's Inn House
127 Clerkenwell Road
London EC1R 5DB
United Kingdom

Trade orders and enquiries:

Telephone: +44 (0)176 760 4972 • Fax: +44 (0)176 760 1640
Email: eurospan@turpin-distribution.com

Individual orders:

www.eurospanbookstore.com/iaea

For further information:

Telephone: +44 (0)207 240 0856 • Fax: +44 (0)207 379 0609
Email: info@eurospangroup.com • Web site: www.eurospangroup.com

Orders for both priced and unpriced publications may be addressed directly to:

Marketing and Sales Unit
International Atomic Energy Agency
Vienna International Centre, PO Box 100, 1400 Vienna, Austria
Telephone: +43 1 2600 22529 or 22530 • Fax: +43 1 26007 22529
Email: sales.publications@iaea.org • Web site: www.iaea.org/publications

22-03143E-T